Women in Chemistry

Their Changing Roles from Alchemical Times to the Mid-Twentieth Century

Marelene Rayner-Canham
Geoffrey Rayner-Canham
Sir Wilfred Grenfell College

History of Modern Chemical Sciences

Anthony S. Travis, SERIES EDITOR
Hebrew University, Jerusalem
O. Theodor Benfey, ASSOCIATE SERIES EDITOR
Chemical Heritage Foundation

Chemical Heritage Foundation

Library of Congress Cataloging-in-Publication Data

Rayner-Canham, Marelene F.
Women in chemistry: their changing roles from alchemical times to the mid-twentieth century/Marelene Rayner-Canham, Geoff Rayner-Canham

   p. cm. — (History of modern chemical sciences, ISSN 1069–2452)

   Includes bibliographical references (p.   –   ) and index.

   ISBN 0-941901-27-0   (acid-free paper)

   1. Women in chemistry—History. 2. Women chemists—Biography.

   I. Rayner-Canham, Geoffrey. II. Title. III. Series.

QD20.R39   1998
540'.82—dc21
98–3890
CIP

The paper used in this publication meets the minimum requirements of American National Standard for Information Sciences—Permanence of Paper for Printed Library Materials, ANSI Z39.48–1984.

For information about CHF publications write
Chemical Heritage Foundation
315 Chestnut Street
Philadelphia, PA 19106-2702, USA
Fax (215) 925-1954

About the Authors

Marelene F. Rayner-Canham is a Laboratory Instructor in Physics at the Sir Wilfred Grenfell College, Memorial University of Newfoundland, Canada. She graduated with a B.Sc. in General Science from the University of Waterloo, Canada. Over the past 10 years, she has been involved in researching and writing works on the history of women in science. Her publications have included *Harriet Brooks: Pioneer Nuclear Scientist*, a biography of one of Ernest Rutherford's first research students, and *A Devotion to Their Science: Pioneer Women of Radioactivity*, a compilation of the lives and work of 23 women researchers in that field in the early part of the twentieth century.

Photographs by: Don Meiwald of Sir Wilfred Grenfell College.

Geoffrey W. Rayner-Canham is a Professor of Chemistry at the Sir Wilfred Grenfell College. He obtained a B.Sc. in chemistry and a Ph.D. in inorganic chemistry from Imperial College, London University, England. For many years, his interests have focused on chemical education and the history of science. As well as collaborating with his spouse on projects in the history of women in science, he has written the college text *Descriptive Inorganic Chemistry*. For his contributions to chemical education, he received the Polysar Award from the Chemical Institute of Canada in 1980 and the Catalyst Award (Canadian Region) from the Chemical Manufacturers Association in 1985.

Contents

Preface

Women have been active participants in the chemical sciences since the beginning of recorded history. Yet books and articles on the history of chemistry rarely make mention of any women apart from the required obeisance to Marie Curie, who by default becomes lodged in the consciousness as the one-and-only woman in the history of chemistry. But this is a travesty of justice, for there *were* other women chemists, and they, too, deserve their place in the "chemical" sun. It is the main purpose of our book to right this wrong and to do so in a volume that will be useful to all those interested in the history of women in chemistry: chemical educators, practicing chemists, historians of chemistry, and participants in women's studies.

Historian scientists, that is, writers of history who are primarily scientists, tend to see the history of science in terms of individuals and their life and contributions. This approach is certainly important, but it is crucial in discussions of women chemists that we also provide the context required by science historians. It is only by looking at the social and historical circumstances that we can understand the forces that have circumscribed women's roles in the progress of chemistry. To this end, although this book details the life and accomplishments of individual women chemists, these biographies are contextualized in every chapter by introductory remarks devoted to the cultural framework of the period. In turn, a significant proportion of each biography shows how the cultural framework influenced a woman chemist's educational opportunities, research direction, scientific accomplishments, and interaction with the male scientific establishment. We have highlighted the major roles of mentors in the progress of most of the women chemists and, where possible, identified links between specific women's lives.

The biographical accounts are intended as a readable overview of each person's life and work, and for this reason, a lengthy bibliography is provided with each chapter for those readers who wish to delve more deeply. Unfortunately, many of the women left little in the way of personal records, and the accounts of their lives must be less complete than we would have liked them to be. Also, this compilation makes no pretense to be a comprehensive collection of women chemists. Such biobibliographic compendia are useful as reference works, but we intend our book to be a thought-provoking introduction to this important topic. On the other hand, we have not restricted ourselves to the "Great Women of Chemistry". Our selection includes the accounts of many very talented women who made significant contributions, but who were thwarted from demonstrating their full ability by the prejudices of their time. We have endeavored to choose the women chemists on the basis of importance and interest; thus the earlier chapters are dominated by accounts of European women while the later chapters have a greater proportion of North American women. If the reader can identify additional women chemists who may fit within the scope of this work, we would be most grateful for details.

The chapter contents are organized according to themes; hence, the time periods sometimes overlap. As well, we have no firm cutoff date in our historical survey, though the vast majority of the individuals were born before 1900. However, if the inclusion of an individual from a later time frame made historical sense, then that person has been included. The overall aim was to conclude with the 1950s, the period that marked the end of the first large influx of women into chemistry and the beginning of the contemporary "second wave" of women chemists.

We must thank Ms. Elizabeth Behrens, Associate University Librarian, Sir Wilfred Grenfell College (SWGC), Memorial University of Newfoundland, for her help in tracking down the many obscure literature sources. Acknowledgment must be given to Dr. Katy Bindon, Principal, SWGC, for approval of a partial relief of teaching (for G. W. Rayner-Canham) for the completion of the manuscript; to Dr. David Waddington for his hospitality during a sabbatical leave at the University of York, England, a base from which some of the research was undertaken; and to New College of the University of South Florida, Sarasota, for a summer research scholar appointment (for G. W. Rayner-Canham), during which some of the chapters were drafted.

Marelene F. Rayner-Canham
Geoffrey W. Rayner-Canham
Sir Wilfred Grenfell College
Memorial University of Newfoundland
Corner Brook
Newfoundland A2H 6P9, Canada

Chapter 1

From the Earliest Times to the Scientific Revolution

The Earliest Women Chemists—Women Alchemists of the Classical Period—The Chinese Women Alchemists—The Dark Ages and the Renaissance—Medieval Alchemy—Biographies: Marie le Jars de Gourney (1565–1645); Marie Meudrac (mid-1600s)—The End of Alchemy

The Earliest Women Chemists

Women have been involved in the practice of chemistry since the earliest recorded times. If we define chemistry as the use of chemical equipment and processes,[1] then it can be argued that the first chemists we can identify by name were two women. These were Tapputi-Belatekallim and (—)-ninu (the first half of whose name has been lost), and both lived in Mesopotamia some time about 1200 B.C. during the period of the Babylonian civilization.[2] According to clay tablets, these women were perfumeresses who obtained the essences from plant sources by means of extraction and distillation procedures.[3] Tapputi, the female overseer of the Royal Palace, was in charge of the manufacture of perfume products, and she had worked out the steps of preparation by her own methods. It is not surprising that these early records point to women as the first chemists, because the equipment often derived from culinary items, and the whole concept of devising systematic and quantitative extraction procedures resembles that of a cooking recipe.

The Egyptian dynastic society (3000 B.C. to 300 B.C.) followed the decline of the Babylonian civilization. In both of these societies,

1

women were largely treated as equals, and in fact, Egypt was ruled by several female pharaohs. Though we do not know of any specifically chemical tasks, Egyptian women of the time were involved in applied chemistry, such as beer making and the preparation of medicinal compounds.[4]

With the rise of the Greek civilization (about 700 B.C.), two changes occurred. First, the Greeks sought philosophical explanations for chemical processes. This mixture of practical chemistry and mystic philosophy we now refer to as alchemy.[5] Instead of the straightforward "recipe" instructions of the Babylonians, we now find obscurity and religious symbolism interwoven in the procedures of the Greeks. The second change relates to the more subservient role of women in Greek society compared with that in Babylonia and Egypt. For example, upper-class Greek women were expected to stay within the women's quarters of the house and to send servants for any outside tasks.[6] By law, women were banned from attending public gatherings. Thus it is not surprising that, with few exceptions, such as Aspasia of Miletus,[7] Arete of Cyrene, and Artemisia of Caria,[8] women were excluded from the philosophical discussions that resulted in the brilliance of Greek mathematics and physical science. As a result, there is no evidence of any women alchemists during the "golden age" of Greece.

Although women were more liberated in the later Roman civilization, the Romans focused on applied technology, and the pure sciences were little developed. Nevertheless, in the period of the later Roman Empire, a number of women made major contributions to alchemy. But it was in Alexandria, not Rome, that these individuals are believed to have lived.

Women Alchemists of the Classical Period

It was the city of Alexandria in the Nile delta of Egypt that served as the focus of knowledge for much of antiquity.[9] Following the founding of the city by Alexander the Great in about 330 B.C., the Great Library was constructed. At its peak, the library amassed nearly 700,000 volumes. From the beginning, the rulers of Egypt recruited some of the foremost scholars of the time, such as the geometrician Euclid, to what was called the Museum. These scholars, in turn, attracted students from the entire Western civilized world, making Alexandria a center of learning for a period of 700 years.

During the early centuries A.D., Alexandria was the home of the most famous woman alchemist, Maria Hebraea, also known as Mary the Jewess. As none of her original writings survived, we know of Maria Hebraea only through the writings of many later alchemists, particularly those of Zosimos of Panopolis. Zosimos, one of the most famous early alchemists, lived in Alexandria during the Roman Period in the late third and early fourth centuries. Though he quoted extensively from the work of Maria with great respect, even awe, he gave no indication of the time or location that she lived. The alchemical historian Raphael Patai considers that she must have lived in Alexandria about the early third century.[10]

Maria Hebraea's greatest impact was in the devising of new and improved alchemical equipment, particularly heating and distilling apparatus. She employed metal, clay, and glass for the constructions, preferring glass vessels where possible as "they see without touching."[10] Her most famous contribution was the *balneum Mariae*, or water bath, a double boiler for gentle heating. Whether she actually invented the water bath or simply improved upon it, her name has survived as a result of the link with the apparatus, for it is known in French as a *bain Marie* and in German as a *Marienbad*. In addition, she devised or improved the distillation equipment of the time, the simple still (called a *kerotakis*) and a complex distillation device (the *tribikos*).

She was obviously familiar with much that we would recognize, for Maria Hebraea describes the properties of mercury as "the deadly poison, since it dissolves gold, and the most injurious of all metals."[11] Also, she gave her name to Mary's Black, the mixed-metal sulfide formed by heating sulfur with a lead–copper alloy. However, she was an alchemist, and she saw chemical change in a manner completely different from that of a modern chemist. For example, she considered that metals were male and female and that they had body, soul, and spirit, which could be revealed by complex alchemical processes. This interpretation of chemical change as some mystical phenomenon meant that most of the writings ascribed to her are wreathed in ponderous, obscure, and often indecipherable language.[12]

Maria Hebraea was not the only woman alchemist of the period.[13] Zosimos had a sister, Theosebeia, who, according to Zosimos, practiced alchemy and possibly coauthored his alchemical encyclopedia. Also mentioned in some alchemical writings are two other women: Cleopatra the Alchemist (Kleopatra Chrisopœia), who is sometimes credited with designing the basic distillation still and who was possibly a contemporary of Maria Hebraea; and Paphnutia the Virgin.

Alexandria was home to at least one other woman of science: Hypatia, the first known woman mathematician. Mention of her is relevant to this account as her demise seemed to mark an end of tolerance toward women in science. Hypatia was born in Alexandria about 350 A.D., the daughter of the author, Theon.[14] Educated by her father, Hypatia became renowned for her work in the field of mathematics and astronomy and for her abilities as a charismatic teacher. In about 415 A.D., she was seized by a crowd of Christian zealots, stripped, and brutally hacked to pieces, and the remains of her body were burned. Though there is dispute about the reason for her murder, it is often argued that she was killed for her activities as an outspoken woman scholar. In any event, her death seemed to mark an end of prominent Western women in science for many hundreds of years.

The Chinese Women Alchemists

It is easy to forget that, in the Far East, an equal, or in many technological respects a greater, civilization had arisen. The Chinese civilization, too, had its alchemy,[15] and among its practitioners were a number of women.[16] The earliest recorded woman alchemist had the family name of Fang, and she lived about the first century B.C. We have details of Fang's life through the writings of the author and alchemist, Ko Hung.[17] According to Ko Hung, Fang came from a scholarly family who were skilled in the alchemical arts. She studied alchemy with one of the Emperor Han Wu Ti's favorite spouses and thus had access to the highest levels of society. Fang was credited with the discovery of how to turn mercury into silver. This process may possibly have been the technique of silver extraction from ores using mercury, the mercury being boiled off to leave the pure silver residue. If this was, in fact, Fang's procedure, it was not widely known in the West until 1570 A.D.[18] Her husband, Chheng Wei, attempted to obtain the secret procedure from her but she refused to provide it, even though he beat and tortured her. Finally, she went insane, possibly the result of mercury poisoning, and killed herself.[19]

More fortunate was Keng Hsien-Seng, who lived about 975 A.D., the daughter of the eminent scholar, Keng Chhien. Keng Hsien-Seng was mentioned in the science writings of Wu Shu, who described Keng as follows:

> When she was young she was intelligent and beautiful and liked reading books. Fond of writing, she sometimes composed praiseworthy poetry, but she was also acquainted with Taoist techniques and could control the spirits. She mastered the art of the yellow and white [alchemy] with

many other strong transformations, mysterious and incomprehensible. No one knew how she acquired all this knowledge.[20]

Keng Hsien-Seng was summoned to the Royal Palace so that the Emperor himself could observe Keng's alchemical procedures. Among these was the ability to transform mercury and "snow" into silver—again, possibly the use of mercury for the extraction of silver from its ores. Among her other chemical skills was the use of a primitive type of Soxhlet process to continuously extract camphor into alcohol.

There have been other women alchemists identified in Chinese literature, including Pao Ku Ko (3rd century A.D.), Li Shao Yun (11th century), Thai Hsuan Nu (uncertain), Sun Pu-Erh (12th century), and Shen Yu Hsiu (15th century). If the names of this many Chinese women alchemists survived, then it seems a reasonable assumption to conclude that many more, whose names have not survived, flourished in the Chinese civilization.

The Dark Ages and the Rise of Christianity

In the West, with the decline of the Roman civilization, it was the Arabic world that kept science alive by original scholarly work and by making copies of manuscripts from Classical times.[21] During this period, we have no evidence of any women scientists.

The recovery of Western civilization is inextricably linked with the rise of Christianity, for it was in the monasteries and nunneries that the skills of reading and writing were nurtured and the writings of the scholars of the past were collected and copied. The German philosopher and botanist Hildegard of Bingen (1098–1179 A.D.)[22] was the most famous of the women scholars, though there were several others, including Hroswitha of Gandersheim (935–1000 A.D.), another German.[23] These powerful abbesses lived at a time of dramatic upsurge in the number of nunneries, which in several parts of Europe now reached the point that they greatly exceeded the number of monasteries. This growth in numbers and power of the nunneries was not welcome to the all-male clerical hierarchy, and it provoked one of the earliest backlashes against women's advancement.[24] Many of the religious orders closed their doors to women and disbanded their nunneries, hence excluding women from the ability to learn to read and write, and particularly to study Latin, the language in which the scientific texts were written. The remaining religious orders for women became orders of charity rather than learning.[25] Without access to education and without the skills in Latin, the world of science became closed to women.

When the earliest Western universities were founded, such as the University of Paris in 1150, women were still excluded from academic education. This was not surprising, for the universities came from monastic roots. As the French historian Achille Luchaire commented, "The university was a brotherhood almost entirely composed of clerics; masters and students had the tonsure; collectively they constituted a church institution.... Universities were ecclesiastical associations and were organized accordingly."[26]

It is of relevance that several hundred years later, in 1561, when the Anglican Church abolished celibacy for clerics, Queen Elizabeth I issued an order that academic celibacy be continued in Britain, and it was not until 1882 that the ban on married Fellows was lifted at Oxford and Cambridge Universities. To emphasize her objection to women's education, Elizabeth reinstated the complete ban on women from universities.[27]

The requirement of academic celibacy had a secondary effect on the ability of a medieval girl to gain an education, as Margaret Wertheim has noted:

> Because medieval professors could not have families of their own (not within the legal bond of wedlock anyway), they too were officially childless. Thus any potential medieval Theon could not have had a daughter like Hypatia upon whom he could have lavished his knowledge. One outcome of the policy of academic celibacy was thereby to disenfranchise women from *indirect* access to academic learning as well.[28]

Thus when we look back over the history of Western science, it is no wonder that women played only peripheral roles until the nineteenth century.[29]

Medieval Alchemy

Though excluded from formal education, several women from the period are known to have practiced alchemy. Little is known about most of these women. They were probably obsessed by secrecy, in large part necessitated by the fear of being branded a witch as a result of their mystical alchemical practices. The fear of witchcraft in medieval society resulted in the deaths of enormous numbers of women, some of whom were the wise women of the communities, such as alchemists, herbalists, and midwives.[30] A particularly relevant account is that of the first recorded woman mineralogist and mining engineer, Martine de Bertereau du Châtelet. More commonly known by her married name of Baroness de Beausoleil, she tried to practice "modern" science in France during the 1600s. In his book *Woman in Science*, first published in 1913, H. J. Mozans described her two published works:

> In these two productions Mme. de Beausoleil treats of the science of mining, the different kinds of mines, the assaying of ores and the divers method[s] of smelting them, as well as the general principles of metallurgy, as then understood. But unlike most of her contemporaries, this enlightened woman had no patience with those who believed that the earth's hidden treasures could not be discovered without recourse to magic or the aid of demons. She was unsparing in her ridicule of those who had faith in the existence of gnomes and kobolds....[31]

It is thus particularly ironic that, in 1642, de Beausoleil was imprisoned for witchcraft and died in jail that same year.[32]

One of the most intriguing examples of a woman alchemist was the seventeenth century aristocrat Mary Sydney, wife of the Earl of Pembroke, who was noted as being "a great chymist, and spent yearly a great deale in that study."[33] However, women alchemists seem to have become most prominent in France, and we know of three in particular: Perrenelle Lethas, Marie le Jars de Gourney, and Marie Meudrac. Lethas, wife of Nicholas Flammel, an affluent scribe, lived in fourteenth century Paris. The pair became interested in alchemy, and in 1382, after years of work, they were able to transform mercury into silver, possibly by reduction of the silver ore followed by distillation of the mercury. Later the same year, Flammel claimed that they had obtained a gold more pliable than common gold (possibly a mercury–gold amalgam): "I may speak it with truth, I have made it three times, with the help of Perrenelle, who understood it as well as I because she helped me with my operations, and without doubt, if she would have enterprised to have done it alone, she had attained the end and perfection thereof."[34] Perrenelle Lethas died in 1397.

Up to this point, we have focused on the societal context in which the women alchemists existed, but we need to balance context with an examination of individual lives. Hence we will now look in more detail at two other French women alchemists, Marie le Jars de Gourney and Marie Meudrac. Unfortunately, the biographies in these early chapters have to be somewhat sketchy as the individuals left little record of their existence. In later chapters, the accounts become better balanced between life and work, though there are frustrating gaps in the historical record for almost all of the women documented. Throughout this book, we focus on such themes as the barriers to formal education, the encouragement of mentors and other interpersonal interactions, the key research work, and the recognition (or lack of) for their contributions. Whenever possible, we provide contemporary quotations to try to give a real glimpse into the personality of each woman chemist. Above all, the most striking feature of the biographical accounts is the determination of tenacious women to pursue their almost-obsessive enthusiasm for chemistry regardless of the challenges and obstacles.

Biographies

Marie le Jars de Gourney (1565–1645)

Born in 1565, de Gourney was the daughter of a minor French aristocrat whose mother opposed her obtaining any education.[35] Her rise to fame resulted from a platonic and intellectual friendship with the leading French philosopher and essayist of the time, Michel de Montaigne. After de Montaigne's death, his widow asked de Gourney to prepare and edit his work. Despite de Gourney's lack of formal education, her edited version of his writings was such a critical success that her name became established in literary and political circles in Europe.

This experience convinced her that a career as a professional woman of letters was within her grasp. However, moving to Paris, she soon found herself in financial difficulties. Her interest in alchemy arose from the hope that she could make her fortune by the conversion of common metals into gold. It was probably the alchemist Jean d'Espagnet, a friend of the de Montaigne family, who provided her with a basic knowledge of the subject. Fortunately, she was able to find a friend who arranged for her to have access to the furnace in a glass factory near her residence in Paris. As her investigations continued, the experimental and scientific aspects of alchemical processes began to fascinate her, but it is unclear whether she made any specific contribution to the transition from alchemy to chemistry.

Her many bitter experiences as a woman in a field that was almost exclusively male caused her to formulate some of the earliest feminist arguments. Though Christine de Pizan (1364–1430) had centuries earlier described eloquently the limitations on women in medieval society,[36] de Gourney was the first woman writer to clearly enunciate the equality of men and women except in terms of opportunity. She argued in her two publications, *Égalité des Hommes et des Femmes* and *Grief des Dames*, that such subordination was due not to the will of God or to nature but simply to the ruthless arrogance of men who had restricted the role of women in society. In correspondence with one of her followers, Anna Maria van Shurman,[37] she added that if women were to further their cause, they must turn to the opportunities that science offered them. In her later years, she was recognized as one of Europe's leading intellects, and her death in 1645 was widely mourned.

Marie Meudrac (mid-1600s)

One of the last women of alchemy was Marie Meudrac, though we know nothing of her life.[38] This French alchemist wrote a treatise in 1666, *La Chymie charitable et facile, en faveur des dames*,[39] that described the experiments she had performed. Her book straddled the alchemy–chemistry transition in that it referred to the three basic elements—salt, sulfur, and mercury—yet it was free of much of the obscurity found in traditional alchemical works. She explained the contents of the 334-page work as follows:

> I have divided this Book into Six Parts: in the first, I treat principles and operations, vessels, lutes, furnaces, fires, characteristics and weights: in the second, I speak of the properties of simples, of their preparation and of the method of extracting their salts, tinctures, fluids and essences: the third treats of Animals, the fourth of Metals: the fifth treats the method of making compound medicines, with several tested remedies: the sixth is for the Ladies, in which there is a discussion of everything capable of preserving and increasing beauty. I have done my best to explain myself well and to facilitate the operations: I have been very careful not to go beyond my knowledge, and I can assure that everything I teach is true, and that all my remedies have been tested; for which I praise and glorify God.[40]

Like many of the women chemists who would unknowingly follow in her footsteps, Meudrac had trepidations about the reception of her writings:

When I began this little treatise, it was solely for my own
satisfaction and for the purpose of retaining the knowledge I have acquired through long work and through various oft-repeated experiments. I cannot conceal that upon
seeing it completed better than I had dared to hope, I was
tempted to publish it: but if I had reasons for bringing it to
light, I also had reasons for keeping it hidden and for not
exposing it to general criticism. I remained irresolute in
this inner struggle for nearly two years: I objected to
myself that it was not the profession of a lady to teach; that
she should remain silent, listen and learn without display-
ing her own public knowledge; that it is above her station
to offer a work to the public and that a reputation gained
thereby is not ordinarily to her advantage since men
always scorn and blame the product of a woman's wit
[mind]....On the other hand, I flattered myself that I am
not the first lady to have something published; that minds
have no sex and that if the minds of women were culti-
vated like those of men, and that if as much time and
energy were used to instruct the minds of the former, they
would equal those of the latter...[41]

Her book must have gained some recognition, for it was reprinted in
1674 and again in 1711. Unfortunately, nothing more is known about
this fascinating individual.

The End of Alchemy

The first formal recognition of women as scientists came in Italy during the eighteenth century, where in contrast to every other Western
country, women, for a brief period, were able to hold formal academic rank. The two most famous of these women were Laura Bassi
(1711–1778)[42] and Maria Agnesi (1718–1799).[43] Bassi, a child prod-
igy, became a professor at the University of Bologna, publishing
papers on Newtonian physics. By the end of the eighteenth century,
she was renowned throughout Europe as the leading woman scientist
of the day. Among Agnesi's works was *Analytical Institutions*, a compilation of the new mathematics, which included her own original contributions. This book was the most complete mathematical text for
the next fifty years. Her name is identified today with the mathematical function described by the curve known as the "Witch of Agnesi."

During this period, the change began from alchemy, a study
based on the secrecy of the craft, to true chemistry, a science involv-

ing the open interchange of knowledge.[44] The development of modern chemistry came later than that of the other physical sciences, and if any site was to be selected as its birthplace, it would surely be France. It is there that we find in the next chapter the continuation of the role of women in chemistry.

Chapter 2

The Chemist-Assistants of the French Salons

Women and the Scientific Revolution—Biographies: Émilie du Châtelet (1706–1749); Marie Anne Paulze-Lavoisier (1758–1836); Claudine (Poulet) Picardet (1735–1820); Albertine Necker de Saussure (1766–1841)—The End of an Era

Women and the Scientific Revolution

The discoveries of the eighteenth century had a more pronounced effect on chemistry than on any other science, for they marked the end of the pseudoscience of alchemy and the foundation of modern chemistry.[1] There were contributions made by chemists in every part of Europe, such as Carl Scheele of Sweden and Joseph Priestley of England, but it was a group of French intellectuals and scientists who were to found the "New Chemistry".[2] The most prominent of these French chemists were Claude Berthollet, Antoine de Fourcroy, Antoine Lavoisier, and Louis-Bernard Guyton de Morveau.

With the progression of the Scientific Revolution,[3] it might be imagined that a more positive era had dawned for intellectual women, but instead, medical science and logic were to be used against them. A medical textbook published in London in 1659 explained that "...the female, through the cold and moist of their Sex, cannot be endowed with so profound a judgement [as men]; we find indeed that they take with appearance of knowledge in sleigh [simple] and easy matters, but seldom reach any farther than to a sleigh superficial smattering in any deep Science."[4] In 1674 the Cartesian philosopher Nicholas Malebranche "explained" that women are intellectually inferior to men because their "cerebral fiber" is soft and delicate, altogether lacking the solidity and consistency of male fiber.[5]

About the same time, the secretary of the Royal Society announced in 1664 that the intention of the Society was to "raise a Masculine Philosophy...whereby the Mind of Man may be enobled with the knowledge of solid Truths."[6] Though the masculine nature of science was accepted as a fact by its practitioners, there was one notable dissenter, François Poulain de la Barre (1647–1723). His book, *De l'égalité des deux sexes*, which was translated into English as *The Woman as Good as the Man*, spelled out some very treasonous concepts for the period. For example, he stated:

> Why is it then, that we assure ourselves, that Women are less fit for such things than our selves? Sure it is not Chance, but Unavoidable necessity that hinders them from playing their parts. I urge not, that all Women are capable of all Sciences and Employment; that any one is capable of all: No Man pretends to so much; but I only desire, that, considering the Sex in general, we may acknowledge an aptitude in one as well as the other.[7]

Yet the prevailing attitude was that science, by definition, was cold and calculating; hence, the only women that could be admitted were those who forswore traditional feminine values and adopted those of the ideal male. One such convert was Mlle. Dupré, known as "la Cartesiènne", whose great accomplishment was to consider herself incapable of tenderness. Another, Mme. de Grignon, rejected the gift of a dog for her daughter, Pauline, on the grounds that it would destroy their household devoted to rationality.[8]

In spite of the prejudice, society women developed an interest in science, and the "Scientific Lady" became a feature of the age.[9] The first of the scientific ladies was the British socialite, Margaret Cavendish, Duchess of Newcastle (1624–1674).[10] Though her life preceded that of the real scientific age, she developed her own atomic theory, arguing that different substances consisted of atoms of different shapes. She insisted on attending demonstrations by famous scientists, such as Robert Boyle, at the most prestigious scientific institution of the day, the Royal Society.[11] Cavendish was the first woman to be admitted to the Royal Society and the only one until 1945, but her success was a result of her own social position and influence rather than a breakthrough for her gender. Yet even Cavendish knew her position: "It cannot be expected I should write so wisely or wittily as men, being of the effeminate sex, whose brains nature has mixed with the coldest and softest elements."[12]

It was in the literary salons of France[13] that the scientific lady flourished. The salons (or dinner parties, in today's terminology) had first been instituted by Catherine de Vivonne, Madame Rambouillet (1588–1666), as a means of social interaction for court ladies outside of the formal royal gatherings. Over the years, they became more and more the focus of intellectual discourse. The historian E. G. Bodek has commented: "Thus the salon was really an informal university for women—a place where they could exchange ideas, avail themselves of some of the best minds of the time, receive and give criticism, read their own works and hear the works of others, and, in general, pursue in their own way some form of higher education."[14] Of course, only a very small subset of women moved in the right social circles to be included in such gatherings. Although salons developed in many European cities, only in France did science became a particular focus. And, as modern chemistry was to be born in France at this time, a few of these women were to play a role in the "birth".

Biographies

Émilie du Châtelet (1706–1749)

The greatest of all these French women of the salons was Gabrielle-Émilie le Tonnelier de Breteuil, better known by her married name of Madame du Châtelet.[15] She discovered her love of learning while young, mastering six languages by the age of 12. Because she was exceptionally tall and big-boned as a teenager and delighted in intellectual conversation, her parents despaired of finding her a wealthy husband. Loathe to condemn her to convent life, they gave her the best possible education to make her (expected) single life more tolerable. Du Châtelet matured into a vivacious, intelligent woman who was determined to control her own destiny. To this end, she married the Marquis de Châtelet, who was older, wealthy, and, to her delight, mostly absent as a result of his military career.

Bolder than other women of her time, du Châtelet was able to participate in Parisienne café society meetings by dressing as a man (though she fooled no one); it was here that she acquired her knowledge of science and developed her contacts among the French scientists and philosophers. One of her friends was the French philosopher Voltaire, and in 1733 they took up residence together in the

Marquis's chateau at Cirey. The Marquis gave his approval provided that Voltaire renovated and maintained the property during their stay!

At this time, one of the crucial questions in chemistry was the nature of fire, and in 1738 the French Academy of Sciences offered a prize for the best essay on the topic. Voltaire, with the assistance of du Châtelet, performed experiment after experiment to investigate the problem.[16] By choosing only the experiments that supported his hypothesis, Voltaire concluded that fire was a material substance, and his essay to the Academy reflected these views. Du Châtelet, observing his experiments and studying their extensive chemical library, concluded the opposite: that fire and heat were not material and that heat and light were related. Determined to express her own opinion, du Châtelet, at night and in secret, wrote her own dissertation on the nature of fire, keeping awake by soaking her hands in ice water. Neither submission won, but both were published along with the three prizewinners.

Du Châtelet was able to progress so far because of her strength of character and high self-esteem. In a letter to Frederick of Prussia, she commented:

> Judge me for my own merits, or lack of them, but do not look upon me as a mere appendage to this great general or that renowned scholar, this star that shines at the court of France or that famed author. I am in my own right a whole person, responsible to myself alone for all that I am, all that I say, all that I do. It may be that there are metaphysicians and philosophers whose learning is greater than mine, although I have not met them. Yet, they are but frail humans, too, and have their faults; so, when I add the sum total of my graces, I confess that I am inferior to no one.[17]

However, at this particular time, physics and mathematics, not chemistry, occupied the center of the scientific stage. Hence the remainder of du Châtelet's life was spent working in these two fields, as well as coping with the heavy demands of household duties.[18] She is most often remembered for her translation of Isaac Newton's *Principia* into French, to this day the only French translation, and one containing du Châtelet's own commentaries.[19] While she was finishing the work, she became pregnant by the writer and great love of her later life, Jean-François de Saint-Lambert. The pregnancy caused her to intensify her scientific efforts: "I get up at nine, sometimes at eight; I work till three; then I take my coffee; I resume work at four; at ten I stop to eat a morsel alone; I talk till midnight with M. de Voltaire, who comes to supper with me, and at midnight I go to work again, and keep on till five in the morning.... I must do this...or lose the fruit of my labours if I should die in child-bed."[20] Sadly her premonition was remarkably accurate. Having survived three previous births, du Châtelet died six days after the birth of her fourth child, in part because of the lack of care for her health during her frenetic writing.

Marie Anne Paulze-Lavoisier (1758–1836)

The most significant part of the Chemical Revolution was the overthrow of the "phlogiston" theory of oxidation.[21] Until 1780, it was generally accepted that formation of a metal calx oxide from a metal involved the loss of a material called phlogiston. When quantitative measurements showed that there was a gain in weight during oxidation, the immediate conclusion was that phlogiston had to have a negative mass. It was Antoine Lavoisier, sometimes considered the "father of modern chemistry,"[22] who led the battle against the phlogistonists. Arguably the greatest accomplishment of his career was his announcement in 1783 that calx formation (oxidation) involved addition of the newly discovered gas, oxygen, rather than the loss of phlogiston. Yet for most of his career, he relied on his spouse, Marie Anne Pierrette Paulze-Lavoisier, to assist him in his work.

Paulze-Lavoisier was the daughter of Jacques Paulze and Claudine Thoynet; her father obtained much of his income from the running of a tax-collection company, the Ferme-Générale.[23] When she was 12 years old, the elderly, but very influential, Count d'Amerval decided to marry her. In spite of threats of dismissal from his position in the tax company, Jacques Paulze repelled these attempts at a forced marriage of his daughter. However, after the pressure had continued for two years, he suggested to Lavoisier, his colleague in the tax company, that Lavoisier might marry his daughter as a safeguard for her future. Lavoisier agreed to wed Marie, and though the relationship originated through expediency, it proved to be a very successful marriage.

Soon after the wedding, Paulze-Lavoisier became interested in Lavoisier's scientific research, and she quickly mastered skills that enabled her to participate in the laboratory work. Her intellect impressed others; for example, when she was only 17, Lavoisier received a letter from the Portuguese monk–scientist, Joachim de Magalhaens, asking after his "philosophical wife."[24] Subsequently, she received formal tutoring in chemistry from Jean-Baptiste Bucquet and Philippe Gingembre, two of Lavoisier's colleagues.[25] For the rest of the Lavoisiers' lives together, much of the research work was to be a joint effort. The science historian Bernadette Bensaude-Vincent has commented on Paulze-Lavoisier's role: "The development of the antiphlogistonist campaign can be traced through a number of letters emphasizing the prominent role of Madame Lavoisier as a full-time collaborator and propagandist of the new chemistry."[26]

Paulze-Lavoisier studied drawing with the famous artist Louis David, and she became such a skilled engraver that she produced the copperplate illustrations for Lavoisier's classic work *Traité de chimie*. Two engravings by Paulze-Lavoisier are particularly interesting as they show her taking notes while Lavoisier directs an experiment on oxygen intake versus carbon dioxide output using his colleague, Armand Seguin, as a human subject.[27]

Life for the Lavoisiers followed a strict regime. During the week, they rose at 5 A.M., then worked in the laboratory from 6 A.M. to 9 A.M. (in those days of private research, scientists would have their laboratory as part of the house). Following this, Antoine Lavoisier left the house to pursue his business activities relating to tax collection, explosives production, and the Academy of Sciences. Finally, after dinner they returned to work in the laboratory from 7 P.M. to 9 P.M. Certain evenings were devoted to the scientific salons. After dinner, Lavoisier would perform an experiment, or there would be music, while Paulze-Lavoisier would often sketch. Many local scientists attended, and visitors from other countries were welcomed. One of the visitors was Benjamin Franklin, and he developed a close friendship with the Lavoisiers.[28] Franklin particularly prized an oil painting of himself by Paulze-Lavoisier. The routine for Saturdays would be quite different as Lavoisier's young followers would visit them. The morning was taken up by communal experimental work; then, after lunch in the laboratory, the afternoon would be devoted to theoretical discussions.[29]

Languages were another field that Paulze-Lavoisier mastered, particularly English and Latin. Using these linguistic skills, she translated scientific works into French including Kirwan's crucial text *Essay on Phlogiston,* to which she added her own footnotes pointing out errors in the chemistry, though she noted that "the translator has added only a few notes of no great consequence."[30] Despite her self-effacing remarks, her comments were of significance and they indicate that her knowledge of chemistry was far from minimal. Paulze-Lavoisier also translated the works of Priestley, Cavendish, and others for Lavoisier's personal use. This was vital for Lavoisier's work, as he

Marie Paulze-Lavoisier (1758–1836) with Antoine Lavoisier (right) in their laboratory. A painting by Jacques Louis David in 1788. (Courtesy of the Metropolitan Museum of Art.)

was not accomplished in languages and needed her translations to keep up with the research of his foreign contemporaries.

The commonly accepted image of Paulze-Lavoisier is largely based on the description by Grimaux, one of Lavoisier's early biographers, who depicts her character as that of the ideal woman of the nineteenth century.[31] The British traveler and writer Arthur Young visited the Lavoisiers and commented: "Madame Lavoisier, a lively, sensible scientific lady, had prepared a déjeuné Anglois of tea and coffee, but her conversation on Mr. Kirwan's *Essay on Phlogiston*, which she is translating from the English, and on other subjects, which a woman of understanding, that works with her husband in his laboratory, knows how to adorn, was the best repast."[32] Not all visitors were impressed by her intellect. The Lavoisiers entertained Gouverneur Morris in 1789, and in his diary, Morris, who believed in traditional roles for women, commented that "[s]he is tolerably handsome but from her Manner it would seem that she thinks her forte is the Understanding rather than the Person."[33]

During Lavoisier's life, Paulze-Lavoisier was more the assistant, but with his imprisonment in 1793 and execution in 1794,[34] she took on a central role. Her first action was a written public attack (signed Paulze, widow of Lavoisier) on Antoine Dupin, whom she regarded as having a leading role in Lavoisier's prosecution. More relevant to the scientific memory of her late spouse, she organized the publication of his final work, *Mémoires de chimie*.[35] This was a compilation of his papers and those of some colleagues, expounding the principles of the new chemistry. Originally Lavoisier intended an eight-volume work, but at the time of his imprisonment only the pages of the second, most of the first, and part of the fourth volumes had been typeset and printed. Paulze-Lavoisier asked Armand Seguin, who had coauthored some of the papers included in the compilation, to write a preface and help prepare the *Mémoires* for publication. However, she insisted that the preface should contain a virulent attack on the revolutionaries who, in her opinion, had been responsible for Lavoisier's death. This, Seguin refused to do. Subsequently, she took it upon herself to write the preface and arrange for the binding and publication of the work.

The printers for this work were P. S. du Pont and his son E. I. du Pont, who had been great friends of the Lavoisiers since 1776.[36] In fact, the relationship between P. S. du Pont and Paulze-Lavoisier became particularly close. This association had begun when Lavoisier started to spend much of his time at his large country property, while Paulze-Lavoisier stayed in Paris. As a result of this separation, Paulze-

Lavoisier and P. S. du Pont spent increasing amounts of time together, becoming lovers in 1781.[37] This relationship was to continue for ten years.

To finance his new venture into printing, P. S. du Pont had borrowed the funds from Lavoisier. When du Pont himself was imprisoned, he was unable to continue repayments to the widow, and at the same time, by revolutionary decree, Paulze-Lavoisier's assets were seized. Penniless and extremely resentful that both her husband and her father had been sent to the guillotine, Paulze-Lavoisier, who had been imprisoned 65 days herself, turned on du Pont demanding immediate repayment. Du Pont wrote to his son: "She has embittered my life. I owe her, we all owe her, all we can do for her. It seems impossible to give her my friendship again, and it was my delight to give her my unreserved affection. I wish I could forget her. But her husband, who was my friend, entrusted her to my care when he was about to die and long before his death."[38] Paulze-Lavoisier pursued her financial claims against the du Ponts for several years, causing E. I. du Pont's wife to write to her husband in 1799: "Mme. Lavoisier must have taken a vow to do us all the harm she can."[39] But by 1805, the enmity seems to have been settled, probably by agreement on repayment of the outstanding debt, as the du Ponts had moved to the United States and were prospering.[40] As a result of settling the dispute, the printing of the *Mémoires de chimie* was completed that same year and Paulze-Lavoisier sent copies to all of the prominent scientists of the day.

The widow Paulze attracted many suitors, the successful one being Benjamin Thompson, the self-styled Count Rumford.[41] Rumford had fled the United States, having supported the British side during the Revolution, and abandoning his first wife in the process. He was one of the most prominent physicists of the time, and just as Lavoisier's work had overthrown the phlogiston theory of oxidation, so Rumford's work on heat had overthrown the view that heat was a material substance called caloric. Rumford wrote to his daughter, Sally:

> I made the acquaintance of this very amiable woman in Paris, who, I believe, would have no objection in having me for a husband, and who in all respects would be a proper match for me. She is a widow, without children, never having had any, is about my own age, enjoys good health, is very pleasant in society, has a handsome fortune at her disposal, enjoys a most respectable reputation, keeps a good house, which is frequented by all the first

philosophers and men of eminence in the science and literature of the age, or rather of Paris,...She is very clever....[42]

Paulze-Lavoisier and Rumford were married in 1805, but even though
they had known each other for four years and had traveled through
Europe living together, the marriage deteriorated within the first two
months. On Rumford's part, he was a rather conceited, boring individual who was expecting to live well on Paulze-Lavoisier's finances,
while pursuing his researches alone. Paulze-Lavoisier, on the other
hand, was used to entertaining the intellectual élite of Paris, and,
undoubtedly, she anticipated being involved in his research work. On
the first anniversary of their marriage, Rumford wrote to his daughter, commenting: "I am sorry to say that experience only serves to
confirm me in the belief that in character and natural propensities,
Madame de Rumford and myself are totally unalike, and never ought
to have thought of marrying. We are, besides, both too independent,
both in our sentiments and habits of life, to live peaceably together...."[43] For her part, Paulze-Lavoisier described Rumford as "the theoretical liberal" who was "in practice a domestic tyrant."[44]

The disagreements between the two soon became public knowledge in Paris. On one occasion, Paulze-Lavoisier had invited a large
number of guests without obtaining Rumford's approval. He locked
the front gates and removed the key before they arrived. Paulze-
Lavoisier then conversed with the guests through the gates and later
poured boiling water over his prize flowers. The separation, which
took place in 1809, was quite amicable, and Sally Thompson, who had
been receiving regular commentaries from her father over the years,
finally visited Paris and met Paulze-Lavoisier for the first time in 1811.
In spite of the critical comments from her father, Sally considered
Paulze-Lavoisier to be gracious, charming, and of "admirable character." Paulze-Lavoisier never returned to her chemical activities;
instead her later years were filled with charitable work.

Claudine (Poulet) Picardet (1735–1820)

Paulze-Lavoisier was not the only woman who acted as an assistant
to a famous chemist during this period. Although Lavoisier's name is
the only French chemist commonly remembered, his colleague Louis
Bernard Guyton de Morveau was equally renowned at the time.[45] A
convert to Lavoisier's theory of combustion,[46] Guyton made a range
of contributions to chemistry, the most influential being his New System of Nomenclature (often credited to Lavoisier). It was Guyton

who, in 1782, had first suggested a system of nomenclature based on chemical composition,[47] and his terminology, such as -*ite* and -*ate* suffixes, is still used today. Until Guyton's proposals, chemicals had been named after their properties, such as "sugar of lead", or the location of their discovery, such as "Epsom salts", or some other feature that gave no indication of chemical composition.

Guyton acquired a collaborator in Claudine Picardet, about whom little is known.[48] Born Claudine Poulet in Dijon in 1735, she married the elderly General Picardet. Like Paulze-Lavoisier's, her salon was a famous center for scientific and literary discourse. She probably met Guyton through the salon, for Guyton's biographer, Georges Bouchard, wrote "As for Madame Picardet, she held a salon, where all those who prided themselves on their knowledge of science or literature sought the honour of being admitted. She was able to enjoy the fantasy of adding to her reputation of remarkable beauty and keen mind, that of a woman of science."[49]

While Picardet was still married to the General, she began assisting Guyton with his researches. Guyton was obviously proud of his protégée, for he insisted that the British traveler Young should meet this "learned and agreeable lady."[49] Young duly called upon them at home and reported that "Madame Picardet is as agreeable in conversation as she is learned in the closet; a very pleasing unaffected woman;...a treasure to M. Guyton, for she is able and willing to converse with him on chemical subjects, and on any others that tend either to instruct or please."[50] Picardet seems to have participated in the laboratory work of Guyton (though without formal recognition). In addition, she followed in Paulze-Lavoisier's footsteps as a translator of important scientific works from Italian, German, Swedish, and English into French for the use of Guyton and his contemporaries.[51] The British chemist Richard Kirwan commented in a letter to Guyton "...and you are very lucky to have such a lady who wants to translate the papers of M. Scheele. I find her incomparable."[52]

When her husband died in 1796, Picardet moved to an apartment in Guyton's Dijon house. Shortly after, she joined him in his Paris house and they were married in 1798. Bouchard remarked: "She had given her friend [Guyton] financial help when she was rich and married him when she was ruined."[53] Like Paulze-Lavoisier, Picardet had no children; in fact, having few or no children was characteristic of the salon women, for intellectual pursuits and spousal collaboration were the major focus of their lives.[54] Picardet and Guyton had a happy relationship that lasted until his death in 1816. There is no record of her after his death.

Albertine Necker de Saussure (1766–1841)

More directly involved in experimental work was Albertine de Saussure.[55] Her interest in science was encouraged by her father, the well-known geologist Horace Bénédict de Saussure[56], and at the age of 10 she had already started a diary for noting scientific observations. She became an active experimentalist, and during an attempt to prepare oxygen gas she burned her face quite badly. However, after her marriage to Jacques Necker, her scientific activity must have declined, for Guyton wrote that he had revived de Saussure's interest in chemistry during a visit by the Neckers.[57] She visited several famous French chemists of the period, including Lavoisier, Fourcroy, and Berthollet, and in a letter to her father, she described experiments that she had performed in their laboratories.[58]

Saussure, like Paulze-Lavoisier and Picardet, entertained the brightest minds of the period, including Voltaire, with whom she had dined alone at a young age (a scandalous event in the view of contemporary French society).[59] Her greatest achievement was a book on progressive education in which she argued that female students should receive a rigorous academic training, including the study of chemistry—a novel idea in those days.[60] This two-volume work had at least nine French editions, one Swiss, two Belgian, a British translation, an American translation and two separate German translations.[61] Her later third volume of the work, focusing on the life of women, was also published separately[62] and an American translation was produced.[63] Her ideas had a major impact as is evident by books on her contributions to education appearing from 1884 through 1938.[64]

The End of an Era

We have seen that at least four French society women were able to act as assistants when Paris was the focus of chemical discovery. Paulze-Lavoisier, Picardet, and de Saussure were fortunate in finding encouragement from the leading chemists of the time, particularly Lavoisier and Guyton. As they knew each other, they provided mutual encouragement. Though their mentors encouraged their scholarly pursuits, many intellectuals actively opposed the rise of the scientific lady. Given the name *les précieuses* by their foes, they were satirized by a number of writers. For example, Edmond and Jules de Goncourt started their attack with: "Novels disappeared from the dressing-tables of women; only treatises of physics and chemistry appeared on their chiffonières...."[65]

The salon culture survived the Revolution but not the Restoration that followed. In fact, upper-class women were singled out for special criticism, particularly for their pernicious influence through the salons and in their championship of the philosophers and scientists.[66] The new elite (supported by the clergy) demanded a domestic ideology for intellectual women which was to be defined by purity, innocence, obedience, and dependence, while the former virtues of knowledge, rationality, judgment, and independence were considered decadent and corrupting. Some women actively proselytized for this revival of the traditional values of female subservience, arguing that it would set a good example for the lower classes, while others were far more reluctantly drawn into conformity. When Mary Somerville, the famous British scientist, visited Paris in 1817, she remarked that she could find no women intellectuals: "Among all I have met only one pretended to know a little music and that was poor indeed, two drew a little, in language and science I met with none except Mme. Biot."[67] So ended this brief flowering of French women chemists.

Chapter 3

Independent Researchers of the Eighteenth and Nineteenth Centuries

Avenues Open to Women—Biographies: Elizabeth Fulhame (late 1700s); Jane Marcet (1769–1858); Helen Abbott Michael (1857–1904); Agnes Pockels (1862–1935)—The Professionalization of Science

Avenues Open to Women

We saw in the last chapter that, during the rise of modern chemistry, several French women acted as assistants in the laboratory. Over the following centuries, the role of assistant continued to be an avenue for women's access to scientific research.[1] Among the more prominent of these women were Sofia Rudbeck, assistant to, and later wife of, Svante Arrhenius, the pioneer physical chemist[2]; Emilie Wöhler, sister of Friedrich Wöhler, who helped in the earliest extraction of aluminum metal[3]; Julia Hall, who assisted Charles Hall in the electrolytic extraction of aluminum metal[4]; and Léonie Lugan, who helped her spouse, Henri Moissan, with the electrolytic synthesis of fluorine gas.[5] Another assistant was the maid of Jacob Berzelius, Anna Sundstrom.[6] Friedrich Wöhler wrote a letter to Berzelius commenting that Anna's preference "...to wash chemical glassware instead of plates and drinking glasses and [that she] would rather prepare hydrogen sulfide than food, is evidence of her scientific turn of mind. Please convey greetings from one of her former pupils."[7] In this chapter, however, we will focus on women who worked alone, before access to university was possible. Sally Kohlstedt, who has focused on early American women in science, characterizes these independents by their "individual, unselfconscious, and largely unrecognized efforts."[8]

Until the beginning of the twentieth century, science was often practiced as an amateur pursuit, and amateur status was the only scientific avenue open to most women.[9] The fields of biology and astronomy became common choices as a result of the commercial availability of microscopes and telescopes, many of which were constructed by women.[10] In biology, one of the key figures was the Dutch entomologist, Maria Merian (1647–1717),[11] and in North America, Jane Colden Farquer (1724–1766) was a pioneer botanist.[12] The earliest woman astronomer of note was the Polish scientist Maria Cunitz (1610–1664), who produced vitally important corrections to Kepler's *Table of Planetary Motions*, and many women were to follow in her footsteps.[13] Geology, too, had a number of women working in the field,[14] but mathematics was more sparsely represented, the French mathematician Sophie Germain (1776–1831) being the most famous. Germain gained fame for her work on the theory of elastic surfaces and for the proof known as Germain's theorem.[15]

Mathematics, biology, geology, and astronomy were relatively easy to practice for they required little in the way of facilities or expenditures. To perform chemistry, one needed sufficient financial resources to build a private chemistry laboratory. Only the rich could afford this, as Walter Weldon commented in 1825: "[Chemistry] requires such an appropriation of time and property; such a variety of expensive and delicate instruments; such an acquisition of manual dexterity; and so much thought and attention to its successful prosecution, as will necessarily confine the professed pursuit of it to a few professors, and enthusiastic amateurs, whom fortune and opportunity favour."[16] It was an even more difficult task for women than men as few women had any financial resources, a woman's spouse or father usually holding financial control. Thus it is not surprising that of all fields of science, there seem to have been very few recorded independent women chemists during the late eighteenth and first half of the nineteenth centuries. One of these rare women chemists was Elizabeth Fulhame.

Biographies

Elizabeth Fulhame (late 1700s)

Regrettably we know little about Fulhame's life except that she was born about the middle of the eighteenth century and that she was married to the physician, Dr. Thomas Fulhame.[17] Her contribution to chemistry was a book on combustion, *An Essay on Combustion with a View to a New Art of Dying and Painting, wherein the Phlogistic and Antiphlogistic Hypotheses are Proved Erroneous,* published in 1794. Fortunately, the preface of Fulhame's book explains how she developed her interest in the subject: "The possibility of making cloths of gold, silver, and other metals by chymical processes, occurred to me in the year 1780; the project being mentioned to Doctor Fulhame and some friends, was deemed improbable. However, after some time, I had the satisfaction of realizing the idea in some degree by experiment."[18]

She decided to publish her research work in book form to ensure her claim to the discoveries and prevent "prowling plagiary."[18] Also, she appeared hopeful that her technique of producing gold and silver cloth by metal deposition had commercial possibilities and that the book would solidify her claim to the process. She accepted that a chemistry tome authored by a woman was likely to have a negative reception in some quarters, writing in the preface these forthright comments:

> It may appear presuming to *some,* that I should engage in pursuits of this nature; but averse from indolence, and having much leisure, my mind led me to this mode of amusement, which I found entertaining and will I hope be thought inoffensive by the liberal and the learned. But censure is perhaps inevitable; for some are so ignorant, that they grow sullen and silent, and are chilled with horror at the sight of anything that bears the semblance of learning, in whatever shape it may appear; and should the *spectre* appear in the shape of *woman,* the pangs which they suffer are truly dismal.[19]

Her interest was in reduction reactions that led to the deposition of metals. Over the years, she studied the reduction of metal salts of gold, silver, platinum, mercury, copper, and tin, using as reducing agents hydrogen gas, phosphorus, potassium sulfide, hydrogen sulfide, phosphine, charcoal, and light. The realization that metals could be produced by aqueous chemical reduction at room temperature rather than by high-temperature smelting was probably her greatest contribution to chemistry.

In addition, her use of light as a reducing agent for metal salts was the first recorded example of photochemical imaging—the chemical basis of the photographic process. This discovery resulted in the inclusion of her name and work in an 1839 review of the origins of photography by the British scientist Sir John Herschel.[20]

From her references to the work of contemporary chemists, such as Lavoisier, Kirwan, and Scheele, Fulhame was obviously well-educated in chemical principles. She disliked both the phlogiston and antiphlogiston theories of oxidation and reduction, though her conclusions were fairly close to the antiphlogistonist approach of Lavoisier. Her particular concern was that many reductions and oxidations only occurred in the presence of water. For example, she noted that water needed to be present for the combustion of carbon in air. Fulhame concluded that the combustion was a two-part process, the first involving the reaction of carbon with the oxygen of the water to give carbon dioxide and hydrogen gas, followed by reaction of the hydrogen gas with oxygen in the air to give water. Thus she had proposed the novel idea that reactions could require more than one step, and she was also the first to publish the concept of a catalytic process.[21] In the specific reaction of carbon and oxygen, the catalytic role of water has since been established,[22] but Fulhame committed the error of extrapolating her findings and claiming that water was needed for all combustion processes.

Despite her trepidations, Fulhame's work was quite favorably received, one commentary being titled "An Essay on *Combustion*, By a *Lady!*"[23] The prominent French chemist J. F. Coindet wrote a lengthy and favorable review,[24] and Benjamin Thompson, Count Rumford, repeated her work on the reduction of gold salts by light, noting: "This agrees perfectly with the results of similar experiments by the ingenious and lively Mrs. Fulhame. It was on reading her book that I was enduced [sic] to engage in these investigations; and it was by her experiments that most of the foregoing experiments were suggested."[25] However, the Irish chemist William Higgins claimed that Fulhame had ignored his own earlier work. This claim was based on his finding that water was essential to the rusting of iron, though he had made no attempt to generalize the role of water as Fulhame had done. In the preface of his subsequent book, an *Essay on Bleaching*, Higgins remarked:

> About four years ago, a very ingenious pamphlet appeared in the name of Mrs. Fulhame, in which this doctrine of mine respecting the decomposition and recomposition of water has adduced and extended to every species of oxidation, and even to the deoxydation of metals in every degree of heat. I did not think myself warranted when I had written, and much less so now, upon mature deliberation, to apply it in that general way. Had this fair author read my book, and indeed I suppose she did not (having

quoted every other treatise upon the subject), no doubt she would have been candid enough to do me the justice of excepting *me* from the rest of my co-operators in science, when she told them they erred for having overlooked this modification of their doctrine, and also when she adduced it as an original idea of her own.[26]

However, Higgins softened his criticism at the end of the preface:

I now beg leave to assure Mrs. F. before we part, that I read her book with great pleasure, and heartily wish her laudible [sic] example may be followed by the rest of her sex; particularly by those who possess talents and means for making chemical experiments.[26]

Fulhame gained greatest acclaim in the United States. The chemist J. Woodhouse mentioned "[t]he celebrated Mrs. Fulhame, a lady whom I am proud to quote on this occasion…[t]his distinguished lady, who is equally an example to her sex, and an ornament to science."[27] In addition, the Chemical Society of Philadelphia elected her a corresponding member[28] following a Society oration by Thomas P. Smith, who declared that "Mrs. Fulhame has now laid such bold claims to chemistry that we can no longer deny the sex the privilege of participating in this science also."[29]

Fulhame's book was translated into German in 1798,[30] and an American edition appeared in 1810.[31] In the American edition, the editor expressed distress that Fulhame's researches had not received sufficient recognition and suggested that it may have been due to the "…pride of science, revolted at the idea of being taught by a female."[32] He added that "…it may be grating to many, to suppose a female capable of successfully opposing the opinions of some of our fathers in science." Considering the comments by Thompson, Higgins, and Coindet, among others, she did receive recognition, at least in those quarters. Her work was then forgotten until 1903, when the famous British chemist J. W. Mellor rediscovered it and devoted a whole paper to an appreciation of her contributions.[33]

Though Fulhame was wrong in her overemphasis on the role of water in oxidation, she deserves credit particularly for her discovery of photoreduction and the very concept of catalysis. But above all, we would identify her as the first solo woman researcher of modern chemistry. And, unfortunately, it was to be the latter half of the nineteenth century before other women chemists were to make significant contributions.

Jane Marcet (1769–1858)

Starting in the eighteenth century, many women developed an interest in the scientific progress of their time, but lacking access to formal education, they found it difficult to follow scientific arguments. To remedy this problem, a new type of book appeared, one in which contemporary science was explained in simple terms. The first of these, appearing in the 1730s, explained Newton's laws for women readers.[34] However, it was at the beginning of the nineteenth century that popular science books, almost all written by women amateur scientists, really flourished.[35] In fact, from 1704 until 1840 there was even a ladies' periodical which focused on science, philosophy, and mathematics, *The Ladies' Diary: or, The Woman's Almanack, Containing many Delightful and Entertaining Particulars, peculiarly adapted for the Use and Diversion of the Fair-Sex.*[36]

The majority of the books focused on the biological sciences, but one, *Conversations on Chemistry* written by Jane (Haldimand) Marcet, was to open chemistry to women readers. Jane Haldimand was born in London, England, in 1769,[37] the daughter of affluent Swiss parents. When she was 15 years of age, her mother died and Jane had to take charge of the large family. In 1799, she married Alexander Marcet, a Swiss physician who, at the time, held an appointment at Guy's Hospital, London. Her husband was an amateur chemist and a prominent member of London's scientific society. Among their circle of friends were the chemist J. J. Berzelius, the political economist Thomas Malthus, the geologist H. B. de Saussure (father of Albertine de Saussure, whom we discussed in Chapter 2), and the feminist writer Harriet Martineau.

It was not a coincidence that a resident of London should have become interested in chemistry. The Chemical Revolution started in France, but with the end of the Revolutionary period and the execution of many of the eminent French scientists,[38] the focus of chemical discovery had moved to Britain.[39] This was the period of the great British chemists such as John Dalton, Joseph Priestley, and Humphrey Davy. Davy gave a series of public lectures in London, England, that were attended by members of the upper-class society. He was delighted by the large proportion of women in the audience, which included Marcet, though he made it plain that, in his view, women should absorb scientific knowledge and transmit it to their offspring, but certainly not practice it.[40] Actually, this commonly held view was quite progressive for the period, for the argument was made that these learned women would be able to have a companionate marriage rather than occupy the traditional subordinate role.[41]

Having had difficulty understanding some of the lectures, Marcet performed her own experiments. Finding this experience gave her a deeper understanding of chemistry, she decided to write an introductory textbook accompanied by experimental work so that others could understand the subject.[42] Like Marie Meudrac and Elizabeth Fulhame before her, Marcet was concerned about the reception of her efforts, noting in the preface: "In writing these pages, the author was more than once checked in her progress, by the apprehension that such an attempt might be considered by some, either as unsuited to the ordinary pursuits of her sex, or ill-justified by her own imperfect knowledge of the subject."[43] To illustrate the concern for her reputation, all the early additions were published anonymously, only those from the 13th British edition (1837) onward bearing her name as author.

The book was written in a conversational style, as its name, *Conversations on Chemistry*, indicated. The style had been popularized by the biologist Maria Jacson (1755–1829) in her book *Botanical Dialogues: Between Hortensia and her Four Children*, published in 1797.[44] Marcet considered such a style was particularly appropriate for the female audience: "Hence it was natural to infer, that familiar conversation was, in studies of this kind, a most useful auxiliary source of information; and more especially for the female sex, whose education is seldom calculated to prepare their minds for abstract ideas, or scientific language."[45] In Marcet's book, the conversations themselves consisted of a series of discourses by a Mrs. B and two students: Emily, who is serious and hardworking, and Caroline, who is more spontaneous and imaginative. Caroline's initial lack of enthusiasm is summed up by her first statement in the book: "To confess the truth, Mrs. B., I am not disposed to form a very favorable view of chemistry, nor do expect to derive much entertainment from it."[46]

The characters are sketched only superficially: the two students are wealthy, well-educated town dwellers between 13 and 15 years of age, while Mrs. B. is identified as Mrs. Bryan.[47] A prominent scientific author of the time was Margaret Bryan, and it is possible that the anonymous Marcet wished to lay a false trail as to the authorship of *Conversations on Chemistry*.

The book was extremely successful, with 18 editions appearing in Britain, the first in 1806 and the last in 1853. Science education for daughters of affluent families had become very fashionable in antebellum America[48], and, as a result, Marcet's book was even more popular in the United States.[49] The first American edition, called *Conversations on Chymistry*, appeared in 1806, and, interestingly, it was printed by the same publisher as the American edition of Elizabeth Fulhame's *Essay on Combustion*. In the later U.S. editions, the American editor often added his own comments, in places criticizing Marcet's views. There were a total of 23 U.S. impressions that appeared between 1806 and 1850. According to some accounts, the total sales in the United States amounted to a phenomenal 160,000 copies.[50] There were at least four French editions published in Paris,[51] a Swiss edition published in Geneva, and an Italian translation of a French edition.[52] Also, Thomas P. Jones, a professor of chemistry at Columbia College, Washington, wrote a text entitled *New Conversations on Chemistry*, which he openly admitted was closely based on Marcet's book. Jones's book, which lacked the humor and personal commentary of Marcet's work, appeared in 12 editions between 1831 and 1848.

Marcet's book had a major role in initiating Michael Faraday's career in science. Faraday, at the time, was a bookbinder's apprentice. It was reading *Conversations on Chemistry* while he was binding it that provided his initial chemical knowledge[53] and helped persuade him that science was his true vocation. As Faraday himself remarked:

> So when I questioned Mrs. Marcet's little book by such experiments as I could find means to perform, and found it true to the facts as I could understand them, I felt that I had got hold of an anchor in chemical knowledge, and clung fast to it. Hence my deep veneration for Mrs. Marcet.... You may imagine my delight when I came to know Mrs. Marcet personally; how often I cast my thoughts backwards, delighting to connect the past and the present; how often, when sending a paper to her as a thank-offering, I thought of my first instructress, and such like thoughts will remain with me.[54]

In fact, one of the great attractions of *Conversations on Chemistry* was the inclusion in subsequent editions of the latest discoveries that she learned from the famous British scientists. She was particularly keen to include the current research of Sir Humphrey Davy, such as the isolation of the alkali metals, potassium and sodium. Some of these reports were quite contentious at the time, as Thomas Cooper acknowledged in the 1818 U.S. edition: "In this publication, she has explained, and explicitly adopted all Sir Humphrey Davy's discoveries and opinions; even when his other contemporaries in chemical investigation, have not yet dared to follow him."[55] At age 71, Marcet wrote to Faraday for news of his latest work:

> Dear Mr. Faraday, I have this morning read in the "Athenaeum" some account of a discovery you announce... respecting the identity of the imponderable agents, heat, light, and electricity; and as I am at this moment correcting the sheets of my "Conversations on Chemistry" for a new edition, might I take the liberty of begging you to inform me where I could obtain a current account of this discovery?[56]

The novelist Maria Edgeworth was another luminary of the time who was influenced by *Conversations on Chemistry*. In fact, her chemical knowledge, acquired by reading Marcet's book, possibly saved the life of Edgeworth's younger sister. The sister had swallowed acid and Maria recalled from the text that milk of magnesia was an effective antidote. Following the incident, Edgeworth wrote of the benefits of studying chemistry by women:

> Chemistry is a science particularly suited to women, suited to their talents and to their situation. Chemistry is not a science of parade, it affords occupation and infinite variety, it demands no bodily strength, it can be pursued in retirement; there is no danger of its inflaming the imagination, because the mind is intent upon realities. The knowledge that is acquired is exact; and the pleasure of the pursuit is a sufficient reward for the labour.[57]

It was particularly important for women writers to show themselves as serious disseminators of scientific knowledge,[58] and in a letter of Edgeworth to a friend, she commented on Marcet's books: "Mrs. Marcet never goes one point beyond what she can vouch for in truth...."[59]

We know that Marcet wrote other books in the *Conversations on...* series, including *Conversations on Vegetable Physiology*,[60] *Conversations on Natural Philosophy,* and *Conversations on Political Economy*. Marcet's *Conversations on Political Economy* was discussed by the French political economist Jean Baptiste, and he commented that Marcet was "the only woman who had written on political economy and shown her superior even to men."[61] Of other anonymous titles, it is difficult to determine how many were authored by her; for example, *Conversations on Botany* is now known to be written by Sarah Mary Fitton. Among Marcet's other writings were a number of children's books. Marcet died in London, England, at the age of 89, and was survived by at least two children.

Helen Abbott Michael (1857–1904)

With Michael, we move forward nearly one hundred years in time, close to the opening of the doors of higher education to women (the subject of the next chapter). However, Michael performed her research while an amateur scientist, only later in life acquiring a formal college education.

Helen Cecilia DeSilva Abbott, born on 23 December 1857 in Philadelphia, was the youngest child of James Abbott and Caroline Montelius.[62] After private schooling, she traveled to Paris in 1878 to study music. While there, she purchased a copy of Helmholtz's work on optics[63] from a second-hand bookstall by the Seine. Upon her return to the United States, she learned, from the family physician, the necessary scientific principles to understand the subject. The book became her favorite reading, and Michael commented that "[t]he horror of my friends and acquaintances at this sudden change in my tastes from Art may be readily imagined."[64] She applied for admission to one of the Women's Medical Colleges[65] as a means of acquiring a broad scientific education. She explained: "This channel seemed the easiest way, as I had not the special preliminary training for entrance to one or two of the colleges then open to women. I did not care to spend the time to secure this entrance knowledge. I looked upon the Women's Medical College (of Philadelphia) as the open sesame to the undiscovered lands."[66] Unfortunately, a fall in her second year caused sufficient injury that she had to withdraw from the College.

In 1883, Michael read of a poisoning of some children who had eaten the roots supposedly of the wild carrot. At the time, to identify a plant from its roots required growing the plant and awaiting the identification of the leaf—by which time the unfortunate victim would have died. This incident generated her interest in plant chemistry. From 1884 to 1887, she worked under Professor Henry Trimble of the Philadelphia College of Pharmacy extracting oils, waxes, and resins from plants. The College trustees were so impressed by her work that they allowed her to lecture to students, the first woman permitted to do so, and paid for a new research laboratory with a designated section for women researchers.

At this period of time, the scientific world was focused on Charles Darwin's work on the origin of species. Michael proposed that evolution might be traced in the chemistry of plant compounds: that is, related plant species would have similar chemical compositions. Her proposal for a chemical taxonomy of plants was to be the great achievement of her career, and she is today recognized as the pioneer of the field.[67] Michael first described her ideas in a paper read in 1886 to the Chemical Section of the Association for the Advancement of Science in Buffalo,[68] and she expanded on her proposal in detailed lectures in Philadelphia and Washington.[69] The *Philadelphia Ledger* reported on her lectures: "The spectacle of a graceful young girl, surrounded by a battery of chemical appliances, and explaining, with the familiarity of an elderly savant, the valuable results of laboratory researches among plants...was interesting from more than one point of view."[70] In the summer of 1887, Michael traveled around Europe. She expressed surprise at the wide recognition her work had gained and at the warmth of the welcome from scientists on that side of the Atlantic.[71]

In the fall of the same year, she returned to the United States and studied chemistry with Professor Arthur Michael at Tufts College, Boston. The following summer they married and settled on the Isle of Wight, England, where Professor Michael equipped a chemistry laboratory for himself, his spouse, and several students. There she performed some research in organic chemistry and published the results. Four years later, they returned to Boston where her husband resumed his position at Tufts College. For a while, Michael remained active in chemistry but she gradually drifted into Boston's social and literary life. One might expect that this would mark the end of her career, but Michael then enrolled at Tufts Medical School, graduating with an M.D. in 1903. She established a private free hospital where she practiced, but, sadly, she died the following year from an infection acquired from one of her patients.

Agnes Pockels (1862–1935)

In the life of Agnes Pockels,[72] we come to the last of the amateur chemists. When she started her research, the universities were still closed to women. Then, when access became available, her parents discouraged her from obtaining a formal education.

Pockels was born in Venice, then part of the Austrian Empire; her parents were Theodor Pockels, a captain in the Royal Austrian Army, and Alwine Becker. Every member of her family suffered ill health on an almost continuous basis, probably a result of living in that malaria-ridden region of north Italy, and her work must be appreciated in the context of Pockels's own recurrent illnesses. In her adult life, her debilitation must have partially resulted from her continuous role as nurse of her sick parents.

Following her father's early discharge from the army on medical grounds, the family settled in Brunswick, Lower Saxony, now part of Germany. Pockels was educated at the Brunswick Municipal School for Girls, recalling later that it was here that her fascination with science commenced: "At that time I had already developed a passionate interest in the natural sciences, especially in physics, and would have liked to become a student, but at that time women were not accepted for higher education and later on, when they started to be accepted, my parents nevertheless asked me not to do so."[73] But this did not stop Pockels, as she described:

> I attempted to continue my education by my own devices, first of all by the use of a small text book by Pouillet-Müller and since 1883 by means of books provided by my brother, Friedrich Pockels, who is three years younger than I and eventually became a professor of physics, but who at that time was a student at Göttingen. However, this type of training did not take me far in respect to the mathematical approach to physics, so that I much regret to have but little knowledge of theoretical matters.[74]

A problem for amateur chemists was the lack of facilities, but Pockels's work in the field of physical chemistry was to be initiated without any equipment or laboratory.[75] As her sister-in-law described, Pockels's scientific work took place

> ...in the kitchen where Agnes, being the daughter of the household, had to do her chores and where it was always nice and warm. In this way, Agnes made her first observations in the field of capillarity. This is really true and no joke or poetic licence: what millions of women see every day without pleasure and are anxious to clean away, i.e., the greasy washing-up water, encouraged this girl to make observations and eventually to...scientific investigations.[76]

Over the following years, she accumulated a substantial quantity of careful research on the nature of surface films. Pockels did not know whether her discoveries were already known as the physicists at Göttingen expressed little interest in her findings. Her savior and mentor was to be the British physicist Lord Rayleigh. Rayleigh had begun studying oil films on water about 1889, and he had published three papers on the subject in early 1890 in a British journal. Pockels would never have discovered this but for a new German science abstract journal, *Naturwissenschaftliche Rundschau*, that was published in Brunswick. We know from Pockels's diary that she opened a subscription to the journal in 1890, in time to read full abstracts of Rayleigh's work.

On 10 January 1891, Pockels courageously wrote Lord Rayleigh:

> My Lord, Will you kindly excuse my venturing to trouble you with a German letter on a scientific subject? Having heard of the fruitful researches carried on by you last year on the hitherto little understood properties of water surfaces, I thought it might interest you to know of my own observations on the subject. For various reasons I am not in a position to publish them in scientific periodicals, and I therefore adopt this means of communicating to you the most important facts.[77]

Pockels then continued, describing the tin trough that she had devised for the study of water surfaces, the beam balance that she had constructed for surface tension measurements, her method of cleaning the water surface, and her studies on the effect of surface films on surface properties. The letter concluded: "I thought I ought not to withhold from you these facts which I have observed, although I am not a professional physicist; and again begging you to excuse my boldness."[77]

Fortunately, Lady Raleigh was fluent in German and rapidly translated the letter into English. Rayleigh replied quickly, asking Pockels to clarify some details of her work, and this she did.[78] Rayleigh could well have ignored this communication from an unknown German woman amateur scientist. Instead, he did all he could to obtain due recognition for Pockels by submitting her work to the prestigious journal *Nature* with a covering letter in which he stated: "I shall be obliged if you can find space for the accompanying translation of an interesting letter which I have received from a German lady, who with very homely appliances has arrived at valuable results respecting the behaviour of contaminated water surfaces."[79] The article was promptly published,[80] establishing Pockels's credentials as a real scientist.

The success of her first publication fired her with enthusiasm, and she continued her research work. In her second paper,[81] she described the method for applying water-insoluble layers to water surfaces by allowing drops of a benzene solution of the compound to evaporate on the surface. This technique is still used today. By this method, Pockels calculated the thickness of the surface film at the point when the surface tension suddenly dropped—what we now recognize as the thickness of the surface monolayer. Over the following years, she published numerous landmark papers in the field of surface science[82] and at last obtained recognition from German scientists, being invited to professional meetings.

Though she published work as late as 1933,[83] the first ten years were the most fruitful. Pockels, herself, explained:

> However, since my time was much in demand for home nursing, I was only rarely able to conduct experiments after 1902.... When my brother died in 1913, the alarums of war and post-war period engulfed me and the *Beiblätter* [*zu den Annalen der Physik*] ceased publication; I was no longer in a position to obtain the relevant literature and in the end I completely lost contact with research in my field, the deterioration in my eyesight and in my health altogether being a contributory factor.[84]

Pockels never did obtain any formal employment. Her sister-in-law added: "During these later years, she led a quiet life as "Auntie Agnes" like many other middle-aged women in Brunswick. She had many acquaintances, and two puzzle-solving societies met in her home.... She herself always lived simply, and kept her thoughts to herself without saying much. The information about her special scientific knowledge was now only noised abroad in whispers."[85] She did, however, receive some recognition of her contributions. In 1931, she was joint recipient of the Laura Leonard Prize[86] of the German Colloid Society for "Quantitative Investigations of the Properties of Surface Layers and Surface Films", and the following year, she was awarded an honorary doctorate from the Carolina-Wilhelmina University of Brunswick. She died in 1935 at the age of 73.

Pockels was to receive everlasting recognition from the work in her second paper. It is usually found that a surface film is compressible as a result of the random orientation of molecules in the monomolecular layer. If a horizontal force is applied to the layer, the layer will contract until all of the molecules are aligned vertically. Beyond this point, now known as the Pockels point, a very much greater force is needed for further compression (that is, stacking of the layers will start to occur). At the Pockels point, the area occupied per molecule represents the cross-sectional area of the molecule.[87]

The Professionalization of Science

During the first half of the nineteenth century, the number of amateur women scientists soared. As an illustration, by 1834 the number of women exceeded that of men at the annual meeting of the British Association for the Advancement of Science.[88] Yet the reputation of one woman overshadowed all of the others—Mary Somerville (1780–1872). At her death, Somerville was hailed as "The Queen of Nineteenth Century Science". Born in Scotland and essentially self-taught, Somerville, a friend of Jane Marcet, pursued a range of scientific interests.[89] With a lack of laboratory facilities, her sole venture into chemistry was a study of light absorption by different materials using the degree of darkening of silver chloride. Somerville performed this project in collaboration with Michael Faraday.[90] Her greatest claim to fame was her book *On the Connexion of the Physical Sciences*.[91] At the time, there was little organization of the different aspects of natural philosophy, as the sciences were then called. Somerville's goal was to draw the disparate threads together and organize the body of knowledge to be known as "science" (and its practicers as "scientists"). The book, acclaimed as a masterpiece of synthesis, appeared in 10 editions and several translations over the next 40 years.

Yet, in another way, Somerville's *Connexion* heralded a turning point in science. By the middle of the nineteenth century, scientific knowledge was becoming too vast to be mastered by one individual, no matter how brilliant. The era of the specialist scientist had arrived, and specialization required formal education. Not until women had access to university education and to the laboratories therein would the number of women in chemistry start to rise.

Chapter 4

The First Generation of Professional Women Chemists

The Fight for Admission to Higher Education—The Opening of the Universities to Women—The Challenges of the Early Years— The Professional Societies—Biographies: Ellen Swallow Richards (1842–1911); Rachel Lloyd (1839–1900); Laura Linton (1853–1915); Ida Freund (1863–1914); Yulya Lermontova (1846–1919); Vera Bogdanovskaia (1868–1897)—The Rise of the New Woman

The Fight for Admission to Higher Education

Up to the middle of the nineteenth century, the lack of access to advanced education was the major factor in the exclusion of women from the professional practice of science. It is important, then, to discuss in some detail the momentous epoch of the mid-nineteenth century when the Western world was indeed "turned upside down" and women gained access to higher learning.[1]

Although we will focus here on women's access to higher education, the first struggle was that for grade school education for girls. In many countries, this step alone was controversial and even when it was not, there was the issue of the most appropriate type of education for women. In particular, the question arose whether the educational content should involve academic subjects or simply focus on those activities that would make girls into better mothers.[2] This point was made concisely in the French Senate as late as the 1880s: "It is not a question of giving them [women] all the knowledge they are capable of acquiring; it is necessary to choose what can be most useful to them, to insist on what best fits their nature of mind and their future status as mothers, and to give them certain studies for the work and occupations of their sex."[3]

Such restrictions on grade school curricula were in themselves a bar to women's access to higher education. For example, French girls' schools did not teach Latin; hence, girls were unable to take the *baccalauréat*, the prerequisite for university entry, which required a knowledge of Latin.[4] Thus as late as 1912, there were fewer French women than women from other countries in French universities.[5] This exclusion mechanism was also present in Germany, where the entrance examination, the *Abitur*, was only offered in boys' schools.[6]

There were three arguments by proponents of a university education for women, particularly a scientific education. These were that women would become better wives and mothers if they had a scientific background; that simple justice and equal opportunity should permit women to enter scientific careers; and that women could perform certain scientific work better than men as a result of their superior patience and manual dexterity.[7]

In fact, the case was made in 1869 by the British feminist Lydia Becker that women would benefit more than men from a training in science:

> Prevalent opinions and customs impose on women so much more monotonous and colourless lives, and deprive them of so much of the natural and healthy excitement enjoyed by the other sex in its free intercourse with the world.... [M]any women might be saved from the evil of the life of intellectual vacuity, to which their present position renders them so peculiarly liable, if they had a thorough training in some branch of science, and the opportunity of carrying it on as a serious pursuit.[8]

Yet many women themselves questioned the role of higher education for their gender, in particular, whether greater knowledge conflicted with the ultimate goal of domesticity.[9] This viewpoint was expressed in 1868 by antifeminist writer Sarah Sewell: "...[P]rofoundly educated women rarely make good wives or mothers. The pride of knowledge does not amalgamate well with the every-day matter of fact rearing of children, and women who have stored their minds with Latin and Greek seldom have much knowledge of pies and puddings, nor do they enjoy the hard and uninteresting work of attending to the wants of little children."[10]

As in earlier times, there was "scientific" evidence that could be used against women's advancement, and now academics turned to the theory of evolution to support their arguments. For example, the sociologist Herbert Spencer had concluded that the difference between the sexes could best be understood in terms of "a somewhat earlier-arrest of individual evolution in women than men."[11] In fact, many scientists of the time had found proof in their research of women's "intellectual inadequacies",[12] and of particular importance, Charles Darwin himself had found "scientific" evidence of female inferiority: "It is generally admitted that with women the powers of intuition, of rapid perception, and perhaps of imitation, are more strongly marked than in man; but some, at least, of these faculties are characteristic of the lower races, and therefore of a past and lower state of civilization."[13]

The medical field, too, added its voice to the undesirability of advanced education for women. Their forceful arguments have been summarized by Ruth Hubbard:

> More effective were the extensive treatises, replete with case histories, that "documented" the drain that menstruation and the maturation of the female reproductive system was said to put on woman's biology and, more importantly, the stress that would fall on these vital capacities if women's intellects were taxed by education. One of the most widely read books of this sort was Edward H. Clarke's *Sex in Education*, published in 1873, which went through seventeen editions in the next thirteen years. Clarke, a former professor at Harvard Medical School and a Fellow of the American Academy of Arts and Sciences, details the histories of many girls whose health, he assures us, was severely damaged by education.[14]

However, it is erroneous to think that Clarke provided the only voice of the medical profession. In the pages of the leading American women's rights publication, *Women's Journal*, distinguished doctors and nonmedical authorities vigorously contested Clarke's views, while Julia Ward Howe edited a compilation of rebuttals to Clarke's claims.[15]

The Opening of the Universities to Women

For countries such as the United States where it was being accepted that women should have access to higher education, there was another fundamental question: should women have completely separate institutions or was coeducation the preferable route?[16] The former view led to the founding of women's colleges. In North America, the first nationally recognized women's college was Vassar (1865), followed by Wellesley and Smith (both in 1875) and Bryn Mawr (1884).[17] But there were major differences in vision for the different colleges; for example, Bryn Mawr, under the leadership of charismatic M. Cary Thomas, aimed toward emulation of the male colleges, while Alice Freeman Palmer of Wellesley favored an educational model that included traditional "womenly" studies.[18] At most women's colleges, science was regarded as an important part of the curriculum; in fact in 1882, M. W. Whitney of Vassar expounded that a scientific education, with its emphasis on logic, simplicity, and reason, was more essential than literary or artistic studies if women were to overcome the image of emotional, illogical creatures.[19]

It was Oberlin College[20] and Antioch College[21] that pioneered the coeducational route. Both of these colleges were founded by Christian social reformers, but though technically coeducational, much of campus life was quite segregated. Pressure from mothers, potential students, civic leaders, and women's rights advocates had been simultaneously opening the doors of traditional mainstream universities. The first of these was the University of Iowa in 1855, followed by Wisconsin (1867); Kansas, Indiana, and Minnesota (1869); and Missouri, Michigan, and California (1870). The only ivy league school to admit women in those early years was Cornell (1872).[22] Many women preferred the coeducational universities, considering the educational experience to be better. Financial and geographical considerations also played a major part in the decision, for the majority of the women's colleges were in the northeastern United States.[23] By 1890, the battle was essentially over, and of the 1,082 U.S. colleges and universities, 43% had become coeducational and 20% were women-only, leaving 37% as men-only.[24]

In Britain, the struggle was to be more protracted.[25] Among the avenues in the 1870s for higher education for women were Girton and Newnham Colleges, affiliates of Cambridge University.[26] The women students were permitted to take the university examinations but were not members of the university, and hence were not eligible for degree status. Even though many women candidates placed unofficially in the top rankings of graduates, it was not until 1948 that women could formally receive degrees from Cambridge University. The University of London was more positive, offering degrees to women in 1878, which could be taken in its coeducational constituents, such as University College and the Royal College of Science, or in the women's colleges, Bedford and Royal Holloway.

In spite of opposition by many prominent university professors to the higher education of women in Germany,[27] this country too followed the trend, with the first woman enrolled at the University of Munich in 1865. However, the obstacle of high school requirements for entry meant that non-German women benefited the most. For example, at the University of Leipzig, during the 1870s, of the 38 women students, 11 were German, 12 British, 10 Russian, 4 American, and 1 Finnish.[28] A few women even acquired doctorates, including Yulya Lermontova and Sophia Kovalevskaia (both of whom are discussed later in this chapter) at the University of Göttingen, and their mutual cousin, Johanna Evreinova, at the University of Leipzig. The accessibility diminished during the 1880s, partially as a result of anti-women protests by the men students. Fortunately, in the German-speaking world, there was always access to the University of Zurich, Switzerland. Zurich had led the way, admitting women in the mid-1860s, and it had become a haven for women from around the world.[29] For example, some of Alfred Werner's classic work in coordination chemistry at the University of Zurich was actually performed by his young research student, Edith Humphrey.[30] Humphrey had traveled from Britain to Zurich to obtain a doctorate in chemistry, which she obtained in 1901.

The situation in Russia was initially very promising, for as early as 1859 women were admitted to university lectures.[31] However, the outbreak of demonstrations by students, at which one woman was arrested, was used as a pretext to restrict women's access to universities.[32] As a result, Russian women had no choice but to travel abroad to obtain a formal university education. In Russia, science was seen to be the solution to the "backwardness" of the country; hence the women who traveled across Europe in search of education were, almost without exception, planning careers in science or medicine.[33] The University of Zurich was usually their goal, and upon arrival, they were astounded to find that they were at the forefront of women's access to university. In fact, of the 203 women at Zurich between 1864 and 1872, 148 were Russian. But for Russian women this haven was to be lost in 1873, when the Russian government announced that the 103 Russian women enrolled at Zurich at that time would have to return to Russia by the end of the year or face permanent exile. This decision was precipitated by the not-totally-unfounded belief that the University of Zurich was a hotbed of political and sexual radicalism, "diseases" that the Russian government did not wish spread at home.

The Challenges of the Early Years

To decide to go to college, then, was a brave act in itself. It required a strong self-image, particularly if the family held to the conventional view that a well-brought-up Victorian or Edwardian girl should stay quietly at home until a suitor appeared on the horizon. Furthermore, in most countries it was an avenue only open to the daughters of the expanding middle class, that rapidly growing business and professional sector of society. The daughters of the poor were simply financially unable to attend university,[34] while the daughters of the upper classes were, for the most part, given an education that would prepare them for their intended life of leisure rather than one that might promote intellectual development.[35] Sharon McGrayne, in her study of women Nobel prize winners in science, noted two factors that were important for success: sympathetic parents and relatives, and family values that stressed education (such as Jewish or Quaker religious philosophies).[36] Although the information is incomplete, one or both of these factors are apparent in the lives of many of the women discussed in this book. Being the eldest daughter (or particularly, an only child) also seems to have favored the pursuit of a university education.[37]

Self-motivation was a major factor in pursuing further education.[38] M. Carey Thomas described her own feelings, which were obviously colored by the perceptions of Edward Clarke and Herbert Spencer: "The passionate desire of the women of my generation for higher education was accompanied through its course by the awful doubt, felt by the women themselves as well as the men, as to whether women as a sex were physically and mentally fit for it.... I was always wondering whether it could be really true, as everyone thought, that boys were cleverer than girls."[39] In addition, to go to college was one of the few avenues (nursing and missionary work being among the others) for a woman to escape the family home without the necessity of marriage.[40]

Going away to college was an exhilarating experience for this first generation. As one student remarked, she and her fellow coeds were happy "...in the glorious conviction that at last, at last, we were afloat on a stream that had a real destination, even though we hardly knew what that destination was."[41] However, along with the joy came the responsibilities of being the pioneer generation. Another student commented that "she bore the weight of formulated womanhood upon her shoulders, although men, even then, were not expected to live to the ideal man."[42] Attitudes from the men students ranged from outright hostility through amused tolerance. At coeducational facilities, women students were often constrained as to where they were allowed to go. In several institutions, women had to enter lecture rooms through a different door from men, and it was common for the lecture room itself to have a separate ladies' row or section. The education historian June Purvis noted that Owens College, Manchester, England, did not allow women students to enter the library: "Ladylike propriety demanded too that women students could not enter the library and ask for a book. Instead, they had to fill in a voucher which was given to a maid-of-all-work, aged about 13, who went to the library with it. If the maid were unsure of the volume, she might have to make the journey ten times."[43]

At Cambridge University, the chemistry laboratories were made coeducational against the wishes of the male lab staff and they made life difficult for the women pioneers. The unidentified writer of the obituary for Girton College student Marion Greenwood commented: "At that time women were rare in scientific laboratories and their presence was by no means generally acceptable—indeed, that is too mild a phrase. Those whose memories go back so far will recollect how unacceptability not infrequently flamed into hostility."[44] Things were little different at Newnham College, as M. D. Ball remembered: "...the lab boys took a delight in leaving some essential bit of apparatus out of our lists so that we had to walk the whole length of the lab to the store to ask for it. An ordeal for some of us, especially as they appeared to be too busy to attend to us for several minutes while we waited at the door."[45]

Hostility was often encountered in dealings with the administrators and faculty. In 1871, the Senate of Heidelberg University, Germany, described the attendance of ladies at academic lectures as "an unsavoury and disturbing phenomenon", and it instructed lecturers "not to tolerate it."[46]

Yet to these women students of the Victorian era, the slights and insults were a small price to pay for the excitement of being in the first assault on the bastions of learning—to be in those "hallowed halls", where studying philosophy or physics was a joy, an end in itself. And even though, by our standards, their activities were severely circumscribed by the need for chaperones, the simple freedom from the societal restrictions at home was, in itself, liberating. The historian J. F. C. Harrison has described this feeling: "For middle class girls the opportunity to have a room of one's own, to be able to organize one's life free from patriarchal dominance, to have cocoa, tea or coffee parties unsupervised, to discuss what one liked with friends, to play games of hockey, and cycle around town—all this was immensely liberating, despite many restrictions and controls imposed by the college authorities."[47]

Finally, graduation did not mark an end to the problems for the woman student. What was she to do with this education? At the time that colleges were opening doors to women, little thought had been given as to what a woman would do with the education.[48] Joyce Antler has shown that most single women graduates actually returned to the parental home, some as "working daughters", while others became "ladies of leisure."[49] A very few—"the independents"—left home to pursue a career. For those graduates between 1865 and 1885, the choices were simply marriage or very low-paid teaching positions and, in fact, the marriage rate of this cohort of students was quite high.[50]

The Professional Societies

The opposition from universities was not the only hurdle facing women chemists. By the end of the nineteenth century, chemistry had become a well-established science, and it had acquired formal structures that, like the other sciences, served to exclude or discourage women.[51] Among these were the American Chemical Society, founded in 1876, and the Chemical Society of London, founded in 1841.

Rachel Bodley (1831–1888)[52] was the first woman member of the American Chemical Society, but she resigned in protest after the Boston meeting of 27 August 1880. The meeting itself was followed by a dinner and "festivities" at which antifemale songs and poems were performed. A subsequent booklet entitled *The Misogynist Dinner of the American Chemical Society* described the events of the evening. The preface of the booklet notes that "[t]he general character of the entertainment...may be gathered from the fact that not only were no ladies invited by the committee, but even when, through a misunderstanding on the part of some prominent members of the society, several brought their wives with them to the dinner, these ladies were refused admission, and actually turned away from the door."[53] After Bodley's resignation, it was 11 more years before another woman member was elected.

In Britain, women were specifically excluded from membership of the Chemical Society. The fight for the admission of women chemists to the Chemical Society commenced in 1880, but, time and time again, motions for entry failed.[54] The prominent chemist Henry Armstrong summed up the feelings of many:

> History proves...the [female] sex to have been lacking in creative and imaginative power.... And it must be so. Throughout the entire period of her existence woman has been man's slave; and if the theory of evolution be in any way correct there is no reason to suppose, I imagine, that she will recover from the mental disabilities which this has entailed upon her within any period which we, for practical purposes, can regard as reasonable.[55]

At the same time, there were some vocal supporters of the rights of women to admission to the Chemical Society. One of these was the editor of the journal *Nature*, who remarked:

> It cannot be denied that women have contributed their fair share of original communications. Indeed, in proportion to their numbers they have shown themselves to be among the most active and successful investigators. The society consents to publish their work, which redounds to its credit. Why, then, should the drones who never have done, and never will do, a stroke of original work in their lives be preferred to them simply because they wear a distinctive dress and are privileged to grow a moustache?[56]

The issue dragged on into the twentieth century. When Marie Curie was elected an Honorary and Foreign Member in 1905, 19 women chemists, led by Ida Smedley MacLean (1877–1944), petitioned for admission. This application, too, was rejected. The moment of victory did not arrive until 1920, and it was appropriate that MacLean, an outstanding researcher in the field of fat metabolism,[57] was the first elected woman Fellow.

In the next sections, we will look at a selection of those dedicated pioneers who led the way into professional chemical careers for women.

Biographies

Ellen Swallow Richards (1842–1911)

One of the best known figures from this early period of university access was Ellen Swallow Richards. Born Ellen Swallow at Dunstable, Massachusetts, the only child of schoolteacher parents, Richards was to be one of the first female professional chemists.[58] Her father had to farm as well as teach to provide enough income, and the young daughter helped with the work in the fields. Determined to provide the best education for her, the family moved to Westford, Massachusetts, where her father opened a village store. She went to school at the Westford Academy, from which she graduated in 1863.

The family then moved to Littleton in the same state, where Richards helped in their new store, taught school, tutored students, and hired herself out as cook, house cleaner, and nurse. When she had saved enough money, she moved to Worcester to attend a school there. Though she enjoyed her independence, the workload and the perceived hopelessness of her ambitions caused her to develop both physical illness and mental depression. It was the opening of the doors of Vassar that proved to be her saving. As she remarked to a friend, "I have been to school a great deal, read quite a little, and so secured quite a little knowledge. Now I am going to Vassar College to get it straightened out and assimilated."[59]

She entered Vassar at the age of 25, and in her second year was promoted to the senior class, enabling her to graduate in 1870 with an A.B. Two faculty members were to have a major influence on her life: Maria Mitchell (1818–1889), Professor of Astronomy, and C. A. Farrar, Head of the Department of Natural Sciences and Mathematics. It was Mitchell who encouraged Richards to study science, and Farrar persuaded her that science should be used to solve practical problems. While at Vassar, she wrote to her parents describing her philosophy: "My aim is now, as it has been for the past ten years, to make myself a true woman, one worthy of the name, and one who will unshrinkingly follow the path which God marks out, one whose aim is to do all of the good that she can in the world, and not be one of those delicate little dolls or the silly fools who make up the bulk of American women, slaves to society and fashion."[60]

Leaving Vassar, she was one of six Americans hired to teach in Argentina but the arrangement collapsed. Richards then applied unsuccessfully for positions as a commercial apprentice chemist. Next, she was given the bold advice to attempt to enter the Massachusetts Institute of Technology (MIT), which had opened as a male-only institution in 1865. In January of 1871, she was admitted to MIT as a special student in chemistry. Her fees were waived, which she assumed was a result of her poverty, and only later did she find out that it was a scheme of the President so that if any protest arose about the admission of a woman (she being the first one), he could deny that she was actually a student.

Though Richards pioneered the entrance of women to MIT, she did not see herself as a proponent of women's rights. In Hunt's biography *The Life of Ellen H. Richards,* Richards is quoted as saying: "Perhaps the fact that I am not a radical or a believer in the all-powerful ballot for women to right all her wrongs and that I do not scorn womanly duties, but claim it a privilege to clean up and sort and supervise the room and sew things, etc., is winning me stronger allies than anything else."[61] In the context of her time, acquiescence to the general views of women's roles was, if nothing else, a wise survival strategy.

In 1873, Richards was awarded an S.B. (B.S.) degree from MIT and an M.A. from Vassar, having submitted to the latter a thesis on the chemical analysis of iron ore. She then spent two years at MIT in graduate work, though she was never formally awarded a doctorate.

While an undergraduate student at MIT, she had helped teach a laboratory chemistry course at Boston's High School for Girls. From her experiences there, Richards became aware of the strong need for practical facilities in science for women. This need led her to fundraising in the Boston community for a Women's Laboratory at MIT that was opened in 1876. This laboratory was mainly used for the training of high school teachers in laboratory science, and it offered practical work in chemical analysis, industrial chemistry, mineralogy, and biology. In 1879, Richards was appointed as an unpaid assistant instructor, a position she held until the closure of the laboratory in 1882 when women were formally admitted to MIT on the same basis as male applicants.

During 1875, she married Robert Richards, the head of the Department of Mining Engineering at MIT. They were accompanied on their honeymoon to Nova Scotia, Canada, by his class of mining engineers so that the class could do their practical field work. The couple, who never had any children, had a very positive marriage, and they were very supportive of each other's endeavors.

The year before she was awarded her undergraduate degree from MIT, Richards had been invited to become a research assistant by Professor William R. Nichols. Nichols had been asked to undertake an extensive study of water pollution by the Massachusetts State Board of Health, and he hired Richards to do the research. This assignment brought her into the field of environmental sanitation. In 1883, MIT opened a separate laboratory for sanitary chemistry, and on the basis of her background acquired with Nichols, Richards was appointed assistant. She participated in the research work on sewage treatment, analyzing many of the 40,000 samples taken as well as administering the research laboratory. As was noted in the final report, "the accuracy of the work and the no less important accuracy of the records were mainly due to Mrs. Richards's great zeal and vigilance."[62] She remained in charge of the laboratory until 1897.

In 1890, MIT established a program in sanitary engineering, and Richards was appointed as instructor in sanitary chemistry, a position that she held until her death. Her course involved the teaching of analytical techniques to students in the sanitary engineering program using the book that she coauthored, *Air, Water and Food from a Sanitary Standpoint*.[63] In addition to her teaching duties, Richards was a consultant for government and industry, testing air, water, and soil samples. The application of chemical analysis to environmental problems pioneered the interdisciplinary field of environmental science.

As a result of working in the family store during her youth, Richards had became interested in the question of food purity. As a result of this concern, in 1886 she wrote a monograph on the subject, *Food Materials and Their Adulterations*.[64] It was this direction, her application of scientific principles to the home environment, that provided her greatest claim to fame. Concerned for the welfare of impoverished families, she developed low-cost, nutritious meals and authored a book expounding her philosophies, *The Chemistry of Cooking and Cleaning: A Manual for Housekeepers*.[65] The first result of these efforts was the opening in 1890 of the New England Kitchen in Boston, where the public could watch the preparation of nutritious meals and then purchase the products. It was not a great success in Boston, as the "Yankee"-based cuisine bore little relationship to the traditional foods of Boston's poor.[66] However, the program was copied in other cities, and it was displayed as the Rumford Kitchen[67] at the 1893 World's Exposition in Chicago. The concept of a scientifically based diet (dietetics) produced interest from schools and hospitals, and the U.S. Department of Agriculture sought her advice for the preparation of nutrition bulletins.

At a series of summer conferences at Lake Placid, New York, organized and chaired by Richards, the study of home science became recognized as the new discipline of home economics. Richards argued vehemently that the improvement of society depended upon the family unit and hence the scientific education of the central figure, the wife and mother. She proposed that women needed to acquire a combination of scientific knowledge, including chemistry, that related to domestic activities, a field that she called "euthenics".[68] As Marilyn Ogilvie has commented, "For the progress of women in science, Richards's applied field was very significant. Little of the opprobrium attached to 'scientific women' appeared in this new area. In studying home economics, women accepted their traditional role in the home while exploring methods of making the home a better place."[69]

Richards died in Boston in 1911. She had been one of the founders in 1882 of the Association of Collegiate Alumnae, an organization that later became the American Association of University Women. Thus it was appropriate that, following her death, the American Association of University Women instituted the Ellen Richards Prize for the woman making the most outstanding contribution of the year to scientific knowledge. In one instance of the interweaving of many of these women's histories, the award was presented in 1913 to Ida Smedley MacLean (a woman chemist mentioned earlier in this chapter) for her work on fat metabolism and fat synthesis.[70] Richards was also one of the charter members of the Naples Table Association for Promoting Scientific Research by Women. This Association, which existed from 1897 to 1933, provided research opportunities for women—a rarity over that time period.[71]

Rachel Lloyd (1839–1900)

Many pioneer women chemists embarked on a chemical career early in life, only to move into other fields later. Lloyd, on the other hand, became a chemist only after the death of her husband, Franklin, an industrial chemist.[72] She was born in Flushing, New York, to the Quaker farming family of Robert Holloway and Abby (Tabor) Holloway, and she was the only one of four children to survive past infancy. Her mother died when she was 5 and her father when she was 12, leaving her to be raised by her father's second wife, Deborah (Smart) Holloway. During her teenage years, she received only a basic education at the Friends School (Quaker) in Flushing. At the age of 20, she married Franklin Lloyd, a Philadelphia chemist. Within the first two years of marriage, she had two children, but both died in infancy, and her husband died shortly after the second child.

After becoming a widow, she spent some time in Europe, returning to take up a career as a teacher, initially teaching at a female seminary in Philadelphia. For six summers between 1875 and 1883 she continued her education by attending the Harvard Summer School, where her main interest was analytical and organic chemistry. In her last few summers at Harvard, she undertook research with the course organizer, the organic chemist Charles F. Mabery. Her work with Mabery on acrylic acid derivatives was of sufficient quality to comprise three papers in the *American Chemical Journal.* In 1883, she accepted the position of professor of chemistry at the School of Pharmacy for Women in Louisville, Kentucky, at the same time becoming instructor at Hampton College, a small women's college in the same city.

Lloyd's long-term ambition was to teach at a university. For this, she realized that she needed a doctorate, which for a woman at the time meant a journey to Zurich. She resigned her positions in Louisville in 1885, traveled to Switzerland, and was accepted as a student under the supervision of Professor V. Merz. Already having had considerable laboratory experience, she was able to complete the research project on the conversion of phenols to aromatic amines at high temperatures in two years. Receiving her degree at the age of 48, she traveled to England in early 1887, spending the first half of that year at the Royal College of Science and the Royal School of Mines in London. It was during her time in London that she accepted an offer of a position at the University of Nebraska.

The University of Nebraska was in desperate need of a second chemist because of a rapidly rising enrollment in chemistry. The original chemist, H. Hudson Nicholson, proposed her name, for he had met Lloyd at the Harvard Summer School where they had both been students. The nominating committee balked at giving her a permanent position, instead initially offering her a one-year term appointment as an acting associate professor of analytical chemistry.

Despite a very heavy teaching load, Lloyd embarked upon a major research program at the Nebraska Agricultural Experiment Station, where she had an appointment as assistant chemist. At the Station, she performed research into the sugar content of sugar beet, a topic of great importance at the time to the farm industry. The sugar beet was a new crop to the United States, the first commercial factories having been built in California in 1879 and 1880. Farmers in Nebraska became interested in sugar beet farming, but before they risked all on this new crop, it was essential for the experimental station to determine whether beet varieties could be identified that had commercial potential in the local soils and climate. This task fell to Lloyd.

During the 1889, 1890, and 1891 seasons, Lloyd, with the help of student assistants, undertook a scientific study of the sucrose content using specific gravity and reduction by Fehling's solution. About 700 complete analyses were performed covering four different varieties of beets and numerous types of soils and climate conditions. Their results showed that, using the appropriate variety and careful farming, sugar beets could be a successful crop in that region. Following from these findings, the sugar beet industry in Nebraska underwent substantial growth, and by 1899, there were already three beet processing factories in the state. This industry has continued to play a major role in Nebraska's agricultural production to the present day.

Lloyd was accepted as a member of the American Chemical Society in 1891, the first woman to join after Rachel Bodley. The same year, her connection with the agricultural station ended when she resigned because of failing health. She continued to have a heavy teaching load at the University of Nebraska where she had been promoted to full professor in 1888, before antifeminist sentiments took hold in academia. She continued to teach there until her resignation due to illness in 1894. She was a popular teacher, the student newspaper devoting a lead editorial to her departure:

> [Dr. Lloyd] has seen develop, largely by her efforts and under her eye, one of the largest chemical laboratories in the West. She has seen her lecture rooms crowded by enthusiastic students of all courses and departments. She leaves in Lincoln many warm, social friends, but it is by the students that her absence will be most keenly felt.... She is one of those instructors who stands not only for a science or a language, but for ideals and all higher culture. We can ill afford to lose one of those, for their name is by no means legion....[73]

She retired to the Philadelphia area, where she had relatives, and died at Beverly, New Jersey, on 7 May 1900. Her early mentor Maberly wrote a moving obituary in which he described her great ability, remarkable energy, and forceful character.[74]

Laura Linton (1853–1915)

Linton's life was the converse of Lloyd's in that she gained success in chemistry early in life but later switched to a career in medicine. Laura Alberta Linton was born in Mahoning County, Ohio, on 8 April 1853, the eldest child of Joseph and Christiana Linton, a Quaker farming family.[75] Laura graduated from Winona Normal School in 1872 and enrolled at the University of Minnesota, Minneapolis, in the same year.

Chemistry became her major interest, and in her senior year she was given the task of analyzing a newly discovered green translucent mineral by two faculty members, Stephen Peckham and Christopher Hall. On the basis of her analysis, they concluded that the mineral was a silicate related to thomsonite. Convinced that it was a unique composition, they named it Lintonite "in honor of Miss Laura A. Linton, a recent graduate of this University to whose patient effort and skill we are indebted for the analysis given in this paper."[76]

After graduation in 1879, Linton taught for a year in the high school in Lake City, Minnesota. During that year, she was contacted by Peckham with an invitation to help him with a Federal Government report on production, technology, and uses of petroleum and its products. She agreed to join him, and they spent two years on the task. As well as general contributions, Linton, being a skilled linguist, translated relevant materials from French and German, and drew the illustrations to accompany the report, acting in a role similar to that of her French predecessors of the previous century (*see* Chapter 2). Peckham effusively thanked Linton in the report itself.

The report was completed in 1882, after which Linton registered at MIT where she studied chemistry for two semesters. She intended to continue her studies there, but the offer of a faculty position at Lombard University, Galesburgh, Illinois, proved irresistible.[77] However, for reasons unknown, she stayed there only one year before returning to a high school position, the head of the science department in Minneapolis Central High School.

About 1894, after 10 years of high school teaching, she returned to research work. The project was a chemical analysis of asphalt samples. It is clear that Peckham had a major role in bringing her back to research, for he supplied the samples and the first year of her research was accomplished in the laboratories of the Union Oil Company, where Peckham was refinery superintendent. These studies were the high point of her chemical career; her work was meticulous and thorough at a time when many contributions to the chemical literature were neither. The research was completed at the University of Michigan, where she was enrolled for the 1895–1896 academic year, perhaps again planning to complete a graduate qualification in chemistry.

At this point, Linton changed her career direction once again, enrolling at the College of Medicine at the University of Minnesota and graduating with an M.D. in 1900. There is no definitive reason known why Linton switched from chemistry, though it may have been associated with a brother and a sister both pursuing successful careers as physicians.

Following completion of her medical studies, Linton joined the staff of the State Mental Hospital in Rochester, Minnesota, where she rose to the position of assistant superintendent in charge of women's wards. She continued to work there until her death on 1 April 1915. During those years, she introduced innovations such as occupational therapy, and the teaching of dietary principles in the nurses' training school.

Linton's life was somewhat of an enigma. Though obviously a talented researcher, she twice discontinued research toward a higher degree and she relinquished the prestigious Conger Professorship of Natural Science at Lombard University in favor of a high school position. Knowing little of her personality and nothing about the circumstances of her appointments, we cannot judge the reasons for her changes of careers. Nevertheless, it is probable that she was able to accomplish more in her later years in the medical field than if she had stayed in chemistry.

Ida Freund (1863–1914)

Just as Ellen Swallow Richards had directed the Women's Chemistry Laboratory at MIT and provided inspiration to an incoming generation of women students, so across the Atlantic, Ida Freund was to play a similar role at the University of Cambridge.[78]

Freund was born in Austria on 5 April 1863, and was raised by her maternal grandparents following the death of her mother.[79] She studied at the State School and the State Training College for Teachers in Vienna. When her grandmother died in 1881, Freund moved to Britain to join her uncle and guardian, the violinist Ludwig Strauss. It was arranged that she would be sent to Girton College, Cambridge, to complete her education, a decision that she bitterly opposed at the time. Her experiences at Cambridge mellowed her attitude, and she became devoted to the place. She completed her undergraduate studies in 1886, attaining a First Class in the final chemistry examination. She accepted a one-year lectureship at the Cambridge Training College for Women and then she was offered a position as demonstrator at Newnham College, the other women's college at Cambridge. She was promoted to lecturer in chemistry in 1890, a position that she held until 1913.

One of her former students, H. Wilson, recalled how Freund "reigned supreme" in the Newnham women's Chemistry Laboratory.[80] Every year, just before the final examination, the Tripos, Freund would summon her chemistry students to do some special study. When the students arrived for the 1907 study session to work on the lives of important chemists, they found that each student had been provided with a large box of chocolates that also contained a written biography of a famous chemist. The following year, when the periodic table was the focus, they found a large periodic table with each element location consisting of an iced cake showing its name and atomic weight in icing, while the group numbers were made of chocolate and the dividing lines were rows of candy sticks.

During her youth, a cycling accident resulted in Freund having a leg amputated, but this did not affect her mobility. M. D. Ball, another Newnham student, recalled:

> Miss Freund was a terror to the first-year student, with her sharp rebuke for thoughtless mistakes. One grew to love her as time went on, though we laughed at her emphatic and odd use of English. Yet how brave she was trundling her crippled and, I am sure, often painful body about in her invalid chair smiling, urging, scolding us along to "zat goal to which we are all travelling which is ze Tripos."[81]

Not only did Freund move freely about campus, but also she became a fervent traveler, "wheelchairing" her way around England, Scotland, Germany, Austria, Switzerland, and Italy.

Apart from her teaching activities, Freund performed research on the theory of solutions which culminated in a substantial paper, and in 1904 she addressed the Cambridge University Chemistry Club on the topic of double salts.[82] Her most renowned work, however, was a chemistry text, *The Study of Chemical Composition*, that remained popular for many years.[83] The historian of chemistry M. M. Pattison Muir commented that her text "is to be classed among the really great works of chemical literature",[84] and the book itself was reprinted in 1968 as a classic in the history of chemistry.[85] Some years later, she also wrote a manual of experimental procedures that could be used to illustrate chemical concepts.[86] For an essay on the early history of the atomic theory, Freund was awarded the University's Gamble Prize. She donated the money to Girton College and the Balfour Laboratory for the purchase of books and scientific apparatus provided that it was spent on "luxuries and not necessities."

Freund was active in feminist causes, being a strong supporter of women's suffrage, and she was one of the three women chemists who first fought for admission to the Chemical Society (the others were Martha A. Whiteley and the previously mentioned Ida Smedley MacLean). In their joint plea on behalf of all British women chemists, they noted:

> Reference to the publications of the Chemical Society shows that during the last thirty years [1873–1903] the names of about 150 women of different nationalities have appeared there as authors or joint authors of some 300

> papers.... Seeing that the Chemical Society recognises the value of the contributions made by women to chemical knowledge by accepting their work for publication, we are encouraged to point out that their work would be greatly facilitated by free access to the chemical literature [in the Society's library] and by the right to attend the meetings of the Society.[87]

Unfortunately, as we saw earlier, their request was in vain.

Freund died on 15 May 1914, and following her passing, the Ida Freund Memorial Fund was instituted at the University of Cambridge to further train women teachers in the physical sciences.

Yulya Lermontova (1846–1919)

The challenges faced by the women discussed above pale into insignificance before those of the Russian chemist, Yulya Vsevolodovna Lermontova.[88] During the middle of the nineteenth century, Russian chemists were at the forefront of the science, particularly Dmitri Mendeléev with his work on the Periodic Table, and Aleksander Butlerov and Vladimir Markovnikov with their contributions to organic chemistry. Chemistry, then, was a logical choice for the scientifically gifted Lermontova. She had been interested in chemistry from childhood, and fortunately her "unladylike" interests were supported by her parents. Her father, the Director of the Moscow Cadet Corps, hired the most talented instructors to provide her with the best possible home education.

At this time, as we mentioned earlier, women were prevented from attending Russian universities. Having been refused admission to the St. Petersburg Agricultural Academy at the age of 22, she realized that to continue her education it was necessary to travel to a country with more progressive educational views. After some time, her parents acquiesced and agreed that she could study at Heidelberg University in Germany, where she would stay with the mathematician Sofia Kovalevskaia[89] and Sofia's husband Vladimir. Kovalevskaia had arrived in Heidelberg in the spring of 1869, only to find that, although the university authorities permitted women to attend some lectures (and even then very reluctantly), they were not allowed to graduate with a degree. Lermontova joined Kovalevskaia in the fall of 1869 and, having been accorded the same chilly reception, obtained permission to attend lectures and to work in the chemical laboratory of Robert Bunsen.

Lermontova studied qualitative and quantitative analysis in Bunsen's laboratory while she pursued her first research work, the separation of the platinum group of metals, a project proposed by Mendeléev. The first few months in Heidelberg were among the happiest of Lermontova's life, and she recalled the "unalloyed delight" of the lighthearted existence in the shared apartment.[90] In 1870, Kovalevskaia's elder sister, Anyuta, and her friend Anna "Zhanna" Evreinova moved into the apartment, and Vladimir moved out. This small group of Russian women became the initial nucleus of what they called their "Heidelberg women's commune". The commune concept flourished among professional Russian women in many academic centers throughout Europe such as Zurich and Paris, and they provided valuable support networks for these women so far from home. Over the following years, many of the brilliant Russian women of the time participated in the Heidelberg commune.[91]

After two years at Heidelberg, Kovalevskaia and Lermontova moved to Berlin. Kovalevskaia was younger than Lermontova by four years, but the former was the dynamic leader of the duo, and to add to her social credibility, she was married. Lermontova recalled that, no longer having the social life of the commune, the two women worked for 16 or more hours every day, Lermontova herself spending nearly all the time in the private laboratory.[92] The historian Beatrice Stillman commented that "...the two women settled down in miserable quarters to an ascetic, monotonous, and isolated existence which seems to have consisted almost exclusively of grinding, unrelieved labor."[93] Fortunately, Kovalevskaia's research supervisor, Karl Weierstrass, took both Kovalevskaia and Lermontova under his wing, having them over for dinner and making sure that they looked after their health.

At Berlin, Lermontova attended the lectures of the great organic chemist A. W. von Hofmann, and she worked in his research laboratories studying diphenene (4,4'-diaminohydrazobenzene), her results providing the first of several research publications. Though Lermontova and Kovalevskaia had both completed enough research work for a doctorate, neither Berlin nor Heidelberg would grant them the degree. Fortunately, the University of Göttingen offered doctorates to foreigners for work completed elsewhere, and Weierstrass was able to arrange for their registration. Even then, the University raised repeated objections to granting degrees to the two women, though it finally capitulated. Thus, in 1874, Lermontova successfully faced an examining committee of the university, one of whose members was the famous Friedrich Wöhler. Both Lermontova and Kovalevskaia were granted doctorates that same year.

Qualified at last, Lermontova returned to Russia, working for a year at the Moscow University with the illustrious V. V. Markovnikov on the synthesis of 1,3-dibromopropane and of glutaric acid. Kovalevskaia returned as well, bringing her daughter, Sofya Vladimirovna (nicknamed Fufa), who was also Lermontova's goddaughter. Lermontova, her sister Sonia, and Kovalevskaia combined their households. This arrangement placed an additional burden on Yulya Lermontova, for she became acting mother of Fufa while Kovalevskaia traveled around Europe looking for an academic position where she could develop her mathematical work.

Becoming ill with typhus, Lermontova was forced to take a year's absence, and after recovery she traveled to St. Petersburg, where she worked with the equally famous A. M. Butlerov, an experience that she called "a true delight."[94] Here she studied the catalyzed reaction between tertiary butyl iodide and isobutylene and she synthesized several new hydrocarbons.

Just as her research reputation was developing, a financial crisis of the Kovalevskaias caused Lermontova to return to Moscow to provide monetary and moral support. Butlerov pleaded with Lermontova to reconsider her departure but without success. He wrote to Markovnikov about Lermontova's desertion, to which Markovnikov replied, "Here the whole reason lies in Sofochka [Sophia] Kovalevskaia. If it were not for her, Lermontova would be in St. Petersburg. That madame, making use of Iulia Vsevolodovna's [Lermontova] kindness, systematically exploits her."[95] Lermontova's departure from St. Petersburg coincided with the restoration of higher education for women in Russia, and she was offered a teaching position in the newly formed "Higher Courses for Women" at St. Petersburg, which she had to reject. Lermontova later claimed that a "boring" laboratory position supervising first-year chemistry students would not have interested her anyway.

The continual postponement of her return to St. Petersburg proved an increasing source of friction with Butlerov, and in 1880 when she finally did return to research, it was once more with Markovnikov. In a way this was fortunate, for it was her next project with Markovnikov that was to give her permanent fame among chemists and petroleum specialists. Markovnikov had developed an interest in petroleum following the development of the vast oil reserves in the Caucasus. He proposed that Lermontova work on the high-pressure cracking of hydrocarbons in the presence of metals. During her research, she developed an apparatus for the continuous processing of petroleum. This was a tremendous technological advance, for until then, crude oil had been processed by batch treatment.

Lermontova had been spending her summer months on the family farm near Moscow. Perhaps realizing that she could never rise in the male-dominated chemical hierarchy, she moved permanently to the farm in 1886, terminating her research career. However, she did not abandon her scientific training, for her agricultural success, particularly in large-scale cheese making, was based on thorough scientific studies. She maintained contact with Kovalevskaia, who moved to Sweden, and after Kovalevskaia's death in 1891, she raised Fufa herself. Lermontova died in December 1919, and though her contributions to chemistry were recognized among her peers, she was never to gain the worldwide recognition of her friend, Sofia Kovalevskaia.

Vera Bogdanovskaia (1868–1897)

Following in Lermontova's footsteps was Vera Estaf'evna Bogdanovskaia. We know much less about Bogdanovskaia, but she is necessary to our history as she represents the first woman who died as a direct result of her research.[96] We do know that in 1890 she traveled to Geneva, where she performed research at the University of Geneva on dibenzylketone for which she obtained a doctoral degree. She returned to Russia in 1892, becoming an assistant inorganic chemist at the St. Petersburg Women's College. At the college, she assisted with the basic chemistry course and also taught a course in stereochemistry.

She was so determined to continue her chemical research that she agreed to a marriage of convenience. The Director of the Izhevsk Military Steel Plant, General Popov, had proposed to her, and Bogdanovskaia accepted on condition that Popov would build her a chemical laboratory in which she could continue her research.

Bogdanovskaia had long been fascinated by the possibility of synthesizing the phosphorus analog of hydrogen cyanide, that is, $H-C\equiv P$. At Geneva, her research supervisor, Professor Groebe, had dissuaded her from choosing its preparation as her thesis project.[97] With her own laboratory, she commenced her research into this compound. On 8 May 1897, during an attempted synthesis, an explosion killed her, and she probably became the first woman to die in the cause of chemistry. In fact, it was not until 1961 that the synthesis of this very reactive and unstable compound was successfully accomplished.[98]

She was well-respected in Russian chemical circles. An obituary written in the *Journal of the Russian Physical-Chemical Society* by one of her former colleagues at the Women's College described a little of her character: "The pleasure of conversation with her was increased by the fact that this woman was thoroughly and comprehensively educated and that she possessed a remarkable lucidity of mind; her papers on chemistry in the literature were remarkable for their clarity of statement.... In these strengths of her personality lie the reasons for her influence on her associates."[99]

The Rise of the New Woman

The first generation, the dedicated pioneers who opened the doors, were students in the period up to about 1890. The second generation, the women students of 1890 to 1915, flooded into the colleges and universities. These were the New Women who knew no boundaries and for whom all was possible.[100] The New Woman had the option of family or career and, as a result, the marriage rate among women graduates declined as the nineteenth century ended and the twentieth century began.[101] This, in turn, provoked an outcry among the American public that the lack of reproduction among educated women would result in the recent immigrants swamping the established Anglo-Saxon American community. It was not until the third generation, after 1915, when marriage *and* a career became an acceptable option.[102]

The rapidity of the growth of women's entry into higher education cannot be overemphasized. Already in 1870, 21% of college students in the United States were women, a proportion that was to climb to 36% by 1890 and peak in 1920 at 47%.[103] However, once entry was gained, the appropriate programs for women became a source of dispute. In California, the discourse was precipitated in 1916 by a woman student at the University of California at Berkeley who, while pipetting, accidentally swallowed a solution containing typhoid bacteria. This accident caused the editor of a newspaper, the *Daily Californian,* to warn about the dangers of feminism and the entry of women into fields that were not suitable. He expressed particular distaste for women working in the sciences:

> Whenever we hear of a college woman who has entered the field of scientific research or is "the only woman in the College of Mechanics" we feel just a little regretful. When her photograph is featured on the front page of the pink-sheet for swallowing microbes or being mixed up in a laboratory explosion, or otherwise making a faux pas in a usually masculine field of endeavor, we always feel a sneaking

hope that the undesired publicity will prove a warning to other ambitious feminine endeavors.[104]

Despite this advice, women continued to choose chemistry-related fields, but as we will see in the following chapters, it was particular areas of chemistry that proved most hospitable.

Chapter 5

Women in Crystallography

Havens for Women—Origins of X-ray Crystallography—The Mentors—Biographies: Kathleen Yardley Lonsdale (1903– 1971); Dorothy Crowfoot Hodgkin (1910–1994); Rosalind Franklin (1920–1958)—Women as Crystallographers

Havens for Women

In the first decades of the twentieth century, professional women chemists were not distributed evenly through all branches of chemistry. Instead, they tended to congregate in certain fields, in particular, crystallography, radioactivity, and biochemistry.[1] This clustering was common in the physical sciences, and the science historian Margaret Rossiter has argued that the fields in which women made up a significant proportion were often the new rapid-growth areas where the demands for personnel were so great that there was less strident objection to the hiring of women.[2]

As we will discuss in this and the following two chapters, there seem to have been two additional reasons that crystallography, radioactivity, and biochemistry were such havens for women. First, they were new, uncharted interdisciplinary areas where success and glory were uncertain. Thus many male researchers avoided such topics, favoring the established research schools instead. The women researchers were less concerned about fame and academic status— the latter being unattainable for women in those days, anyway—and instead it was the pleasure and excitement in the work itself that mattered. This enthusiasm is best summed up by some comments by Vivienne Gornick on the basis of interviews with modern women scientists: "Each of them had wanted to know how the physical world worked, and each of them had found that discovering how things worked through the exercise of her own mental powers gave her an intensity of pleasure and purpose, a sense of reality nothing else could match."[3] In fact, such enthusiasm had to be the basis of their lives, for many were unpaid and those that were paid often lived from one subsistence research grant to another. Formal university rank was a rarity for a woman scientist. Yet the women of those early decades of the twentieth century expressed no open resentment at the situation. In Marcia Bonta's study of the lives of pioneer American botanists, she found this same attitude: "Because most women naturalists believed that the work was all that mattered, they seemed to feel little or no rivalry toward the more powerful males in their fields and were pleased and grateful for whatever help these men gave them."[4]

Second, a point that we will explore in each chapter, the leaders in crystallography, radioactivity, and biochemistry were dynamic young experimenters rather than representatives of the established order and were, for the most part, exceptionally supportive of women in science. This was essential, for a woman had to find a researcher or mentor willing to take on a female student.[5] To explore the situation in the case of crystallography, it is necessary to examine the origins of the subject.

Origins of X-ray Crystallography

X-ray crystallography is one branch of physical science whose origins can be exactly defined.[6] In 1912, the theoretician Max Laue persuaded the experimental physicists Walter Friedrich and Paul Knipping to test his hypothesis that crystals diffract X-rays. The results of this successful experiment were read in a journal by William Henry Bragg (1862–1942), a professor at the University of Leeds, England, and his son, William Lawrence Bragg (1890–1971),[7] then at Cambridge University. They quickly saw the potential for the determination of crystal structures. W. H. Bragg devised and constructed the Bragg X-ray spectrometer for crystal structure analysis, and shortly afterward, W. L. Bragg deduced the relationship between the diffraction pattern and the atomic spacing (Bragg's law). This was the start of X-ray crystallographic methods for the determination of structures.

The First World War interrupted the studies, but with its end, the Braggs decided to divide the crystal world between them. W. H. Bragg chose to work on organic structures and also quartz, while W. L. Bragg was allotted inorganic substances (except quartz). About this time, they both moved: W. H. Bragg to University College, London and then to the Royal Institution, London,[8] while W. L. Bragg obtained a faculty appointment at the University of Manchester, England, a position that he acquired as successor to Ernest Rutherford, who had moved to Cambridge.

The Mentors

The research groups of both Braggs contained remarkable numbers of women.[9] Of the twelve researchers that W. H. Bragg took with him on his move to the Royal Institution in 1923, three were women: I. Ellie Knaggs, G. Mocatta, and Kathleen Lonsdale, and over the following few years, seven more women joined the group: Natalie Allen, Thora Marwick, Lucy Pickett, Helen Gilchrist, Berta Karlik, M. E. Bowland, and C. F. Elam. W. L Bragg's first research student was Lucy Wilson, from Wellesley College, Massachusetts, and he, too, had many other women researchers working with him over the years, including Elsie Firth, Helen Scouloudi, and P. Jones. Maureen Julian, herself a former researcher with Kathleen Lonsdale, has shown that women have contributed throughout the "family tree" of crystallographers, many of the Bragg students (both male and female) themselves taking on women students when they acquired academic positions.[10]

Why was crystallography such an attractive field for women? One viewpoint is that the Braggs and their crystallographic heirs provided a women-friendly environment. Anne Sayre, spouse of crystallographer David Sayre, is convinced that, at least in crystallography, it was the nonaggressive and friendly attitudes of the supervisors that were so vital to the encouragement of women.

> There is something in the ancient history of crystallography that is hard to isolate but nevertheless was there, that I can best describe as modesty. I have often wondered how much the Braggs were responsible for the unaggressive low-key friendly atmosphere that long prevailed in the field (and no longer seems to very much). Somehow the first and second and a few of the third generation crystallographers consistently conveyed an impression of working for pleasure, for the sheer joy of it—the idea of competition didn't seem to emerge very strongly until the 1960s or so. Uncompetitive societies tend to be good for women.[11]

Both Braggs were renowned for their pleasant honest personalities. For W. H. Bragg, this can be illustrated by a letter that he wrote to the Editor of the journal *Nature*:

> I send you a short note which I hope you may see fit to publish in *Nature*. I would like you to know however that my writing it has something to do with an attempt to do a little act of justice to the lady mentioned in the note, Miss Knaggs. She has been working for some time on an extraordinary substance, cyanuric triazide. It is one of the highly explosive nitrogen compounds. In the recent Faraday Society discussion—see *Nature*, May 26—reference was made to her preliminary results without mention of the source from which they had come. It was an accident, of course: there is no question of any unfairness. But this is Miss Knaggs' magnum opus so far and she is naturally disappointed. I have thought I might put matters straight by writing you the short note to which I have referred.[12]

W. L. Bragg's personality can be gleaned from this comment by one of his former students, W. M. Lomer: "Bragg was a gentle man and a gentleman. He never embarrassed anyone and my every contact with him was a pleasure. When he met me showing my fiancée around the little museum...[he] was so very pleasant that to this day my wife will hear nothing against him. And nor will I."[13]

A less charitable view of why women chose crystallography has been taken by Portugal and Cohen, who used the laborious nature of crystallography to explain the number of women in the field:

> Since the high speed computer had not yet been invented, the business of calculating data was a very laborious occupation and smart fellows who could find other things to do would generally do them, unless they were absolutely dedicated to the business of X-ray crystallography. Is it possible that these first class women got to be X-ray crystallographers because they were willing to do this work[?][14]

This parallels the arguments used to account for the very high proportion of women in astronomy during the late nineteenth and early twentieth centuries.[15] Rossiter has shown that the rise in women's participation in astronomy corresponded to the change from the active work of observing (men's work) to the passive role of classifying the thousands of photographic plates (women's work).[16] In fact, it was the first of the women astronomer-assistants, Williamina P.

Fleming, who commented that women's superior patience, perseverance, and method made such activities particularly suitable for women in science.[17] John Lankford and Rickey Slavings, in their studies of women in astronomy, added: "In evaluating women [for astronomy], male scientists tended to focus on their [women's] ability to do routine work. Indeed, some recommendations read as if they were descriptions of machines."[18]

Yet the women described in this chapter did not see their work as dull and routine—to them, the research was the exciting focus of their lives. But, as we mentioned earlier, to enter the field they had to find a supervisor willing to entertain the idea of women students, not necessarily a widespread view in the early part of this century. It was the Braggs who filled this role in the beginnings of X-ray crystallography. This view is supported by Kathleen Lonsdale, one of W. H. Bragg's former students, who commented that the influence of Bragg at the Royal Institution had resulted in the number of women researchers there rising to 20 percent of the total.[19] And it is the life and work of Lonsdale that we will consider initially.

Biographies

Kathleen Yardley Lonsdale (1903–1971)

The first of the women crystallographers to attain prominence was Kathleen Lonsdale (née Yardley).[20] Lonsdale was born near Dublin, Ireland, in 1903, the youngest of 10 children. When 10 years old, her parents separated and her mother took the younger children to live in a small town just east of London, England. She attended a High School for Girls, but as it offered little science, she attended classes in physics, chemistry, and advanced mathematics at the High School for Boys. Though she excelled at almost every academic subject and was a good gymnast, her first love was mathematics. At the age of 16, her exceptional talents resulted in the offer of a place at Bedford College, a small women's college of the University of London.[21] After one year, she switched to physics and, as Lonsdale recalled, "I well remember the opposition I met with from the headmistress of the secondary school that I attended when I told her of my intention to specialize in physics."[22] The headmistress argued that Lonsdale was a fool to think that she would be able to compete in a "man's field". Lonsdale, however, had rejected the alternative of a mathematics degree as she believed mathematics would only lead to a teaching position while physics offered the opportunity of experimental research.[23]

At the age of 19, Lonsdale was awarded an honors B.Sc. degree. W. H. Bragg was one of her B.Sc. oral examiners and, impressed by the fact that Lonsdale had higher university marks than anyone in the previous 10 years, Bragg offered her a position in his laboratory at University College. She accepted his offer, commenting: "He inspired me with his own love of pure science and with his enthusiastic spirit of enquiry and at the same time left me entirely free to follow my own line of research."[24] The following year, she moved with his group to the Royal Institution.

Her master's dissertation, completed in 1924, was on the structure of succinic acid and related molecules. While working on her M.Sc., she constructed a set of 230 space-group tables, mathematical descriptions of the crystal symmetries that became vital tools for crystallographers.[25] Three years later, in 1927, she received the prestigious D.Sc. degree (a higher qualification than a Ph.D.) for her work on ethane derivatives, and that same year of 1927, she married fellow student Thomas Lonsdale. She considered giving up research to become a traditional wife and mother, but Thomas Lonsdale argued that he had not married to obtain a free housekeeper and that she should keep on working. They shared the shopping for food, while she specialized in devising dinners that could be prepared in 30 minutes.

The Lonsdales moved to the city of Leeds where Thomas Lonsdale was working for the British Silk Research Association. Upon her departure from the Royal Institution, she wrote to W. H. Bragg: "I should like to take this opportunity of thanking you again for all the help you have given me in so many ways. I feel sure that it will be difficult to find a place where I shall be as happy in my work as I was at the Davy–Faraday [the Royal Institution Laboratories]."[26]

Upon arrival at the University of Leeds, Kathleen Lonsdale was given research space (probably on the basis of a recommendation by W. H. Bragg), and with a grant from the Royal Society, she was able to purchase an ionization spectrometer. The work at Leeds between 1927 and 1929 on the structure of benzene[27] was to give her international recognition. Auguste Kekulé had proposed in 1865 that benzene had a six-member ring structure and this was generally accepted by chemists,[28] but the question remained as to whether the ring was planar or puckered. The Braggs had already shown that the carbon atoms in diamond could be pictured as forming six-member puckered rings, and benzene was expected to possess the same puckered structure. As benzene was a liquid at room temperature, she used hexamethylbenzene, a solid that could be obtained in large single crystals. Using her spectrometer, a few X-ray photographs taken at the Royal Institution, and some very innovative approaches to structure solving, Lonsdale was able to show that hexamethylbenzene was planar and hence benzene itself was also likely to be planar. Although her results contradicted Bragg's belief in a puckered ring, he was enthusiastic in his praise of her work.

Between 1929 and 1934, she had three children. This was a difficult time for the family, as Kathleen Lonsdale had no formal position and, to make matters worse, in 1930 Thomas Lonsdale's job was terminated. The family returned to London where Thomas Lonsdale obtained an appointment working at the Testing Station of the Ministry of Transport. From 1929 until 1931, Kathleen Lonsdale worked at home on the calculations needed to solve her next crystal structure, that of hexachlorobenzene. To help her find time for the calculations, W. H. Bragg arranged for a small grant from the Royal Institution so that she could pay for a part-time daily home-help.

In 1931, Bragg wrote Lonsdale an enthusiastic letter, telling her that he had obtained funds on her behalf to cover the cost of a full-time home-help so that Lonsdale could return to the Royal Institution. However, when she arrived, she found that all the X-ray equipment was in use. Discovering a large old electromagnet, she began studies on the diamagnetism of aromatic compounds. After completion of this research, she studied the lattice constants of natural and synthetic diamonds. Her precise work on the diamond structure became so renowned that the discoverers of a rare hexagonal form of diamond announced that it would be named lonsdaleite in her honor.[29]

Bragg obtained a grant or fellowship for Lonsdale each year until his death in 1942, after which she was appointed a Dewar Fellow at the Royal Institution with Sir Henry Dale. It was her isolation after Bragg's death that caused her to apply for her first academic appointment. In 1946, at the age of 43, she finally obtained such a position—as Reader in Crystallography at University College where she had started out almost 25 years earlier. Initially, most of her time was filled in the role of editor-in-chief of the *International Tables for X-ray Crystallography*.[30] Three years later, she was promoted to Professor of Chemistry and Head of the Department of Crystallography. Her reputation continued to grow, and she was one of the first two women to be elected a Fellow of the Royal Society, being nominated by W. L. Bragg[31]; in 1966 she was elected the first woman president of the International Union of Crystallography. Among her later work was the study of boron–nitrogen analogs of carbon species, including the identification of a graphite-like form of boron nitride containing alternating boron and nitrogen atoms in arrays of hexagonal rings.[32]

Having lived under the path of the air attacks on London in the First World War and having seen the horrific fireball of a bomb-carrying Zeppelin shot down in flames, Lonsdale became a pacifist. In a letter to A. V. Hill, she expanded on her beliefs:

> It was certainly not the imperative of a religious upbring-ing that made me a pacifist. I like to believe it was com-monsense. I came of a military family and have a naturally pugnacious character, the more violent manifestations of which I have to keep continually under control. My reli-gious teaching as a child was of the orthodox kind that rubber-stamps any war that happens to be going. I don't believe in sitting down and being walked on; I believe in non-violent resistance because I believe it to be the only form of resistance that can be really effective, in that it does not perpetuate the evil it aims at eliminating.[33]

In 1936, both she and her husband became Quakers, and in 1943, she was jailed for a month for refusing to register for war duties. From her personal experiences in jail and as an extension of her beliefs, she became active in prison reform and served on many prison boards. Her indignation at the extensive nuclear testing by the Soviet Union, the United States, and Great Britain caused her to write a book, *Is Peace Possible?*[34], the foreword of which contains her comment that the book was "written in a personal way because I feel a sense of cor-porate guilt and responsibility that scientific knowledge should have been so misused."

In these later years, she relied more and more on Thomas Lons-dale to help with her tremendous amount of correspondence. He had always been very supportive of his famous spouse, and Kathleen Lonsdale identified the need for such a relationship in her comments on women scientists: "For a woman, and especially a married woman with children, to become a first class scientist she must first of all choose, or have chosen, the right husband. He must recognise her problems and be willing to share them. If he is really domesticated, so much the better."[35] In this regard, Kathleen Lonsdale was very fortu-nate. After her death, Thomas Lonsdale wrote to Sir Lawrence Bragg:

> When the apple fell on Newton's head someone gathered it and the other windfalls and made a pie for his dinner, thats [sic] my job now a bit, it always has been. Hilton [a prominent mathematician of the time] told me that a Pro-fessor of Maths is lucky if in his life he has one student who

can see a whole branch of maths as a structure, they know the text book but they don't need it because they go on to build, where the math doesn't yet exist they invent it, for him, Kathleen was that student. Even before we were married I knew she had one of the most powerful intellects of the time.... Most of my working life has been spent in road engineering, I only know enough about her work to realise its importance and value and how fortunate I have been to have been associated with it, 'in getting Newton's dinner'.[36]

Crystallography was her life, and she began a study of the crystalline nature of kidney, bladder, and gall stones when nearly 60 years old.[37] She developed an interest in stones after the Chief Medical Officer of the Salvation Army had told her of small children in hot, dry Third World countries suffering from them.[38] When she fell ill, the first medical diagnosis assumed that she had acquired tropical malaria from her world travels in pursuit of stones and their causes. When cancer of the bone marrow was identified and she was told that her remaining time was short, she began 13-hour work days to complete a book on "Human Stones," the first draft being completed a few weeks before she died in 1971. Ten years later, in recognition of her contributions to crystallography, the then-chemistry building at University College, London, was named the Kathleen Lonsdale Building.

Dorothy Crowfoot Hodgkin (1910–1994)

The most famous of all women crystallographers was Dorothy Crowfoot Hodgkin, the Nobel Prize winner, who determined the crystal structures of penicillin, vitamin B_{12}, and insulin.[39] She was born in Cairo on 12 May 1910, where her father, John Winter Crowfoot, a graduate of Oxford University, supervised Egyptian schools and administered ancient monuments for the British government. Her mother, Grace Mary (Molly) Hood, was a self-taught expert in botany who had written a book on the flora of the Sudan. In 1914, with the outbreak of war, Hodgkin and her two younger sisters were left in England in the care of a nursemaid while their parents returned to the Middle East. Hodgkin believed this background led to her quiet, independent character. Even after the War, the children stayed in England while the parents returned from the Middle East on annual visits.

A Young Dorothy Hodgkin (1910–1994). (Courtesy of Judith A. K. Howard.)

Hodgkin became interested in chemistry at an early age. At 10, she tried growing crystals of copper(II) sulfate and alum, and at 13, when visiting the Sudan, she tried analyzing local minerals. Back in England, she set up a small attic laboratory where she continued her chemical analyses. Of particular importance for her future career, she was given a children's book, *Concerning the Nature of Things* by W. H. Bragg, which discussed crystallography, together with a text on biochemistry, which she read with the help of the *Encyclopedia Britannica*. To enable her to gain a formal knowledge of science, she was given permission to join the boys' science courses in grade school.[40] This experience had the secondary benefit of exposing her to an all-male environment at an early age, so that later in life she seemed quite comfortable in such circumstances.

In 1926, she took the Oxford Senior Local Examination, a national university entrance test, and attained the highest grade for any female student that year. Lacking the Latin and another science required for admission to Oxford, she had to spend a year studying these subjects before she was admitted to Somerville College, Oxford. At this time, her father was a director of the British School of Archaeology in Jerusalem, and she visited him, becoming entranced by the Byzantine mosaic floor patterns. She even considered changing career plans to archaeology—perhaps not such a different career, for as Julian has noted, analyzing mosaic patterns has a strong parallel to her later work on crystal symmetry.[41]

After completing her degree in 1931, she was turned down for several positions, but then J. D. Bernal let her join his group at Cambridge. There she studied early preparations of vitamin B_1, vitamin D, and several of the sex hormones. Bernal was a former student of W. H. Bragg, and he shared the Braggs's philosophy of accepting women researchers.[42] In fact, his first student had been a woman, Nora Martin Wooster, and later women researchers included Helen Megaw, Kate Dornberger-Schiff, and Rosalind Franklin. Bernal throughout the rest of his life was to be a crucial mentor to Hodgkin.[43] Hodgkin described the very pleasant working atmosphere, particularly the convivial lunches: "Every day, one of the group would go and buy fresh bread from Fitzbillies, fruit and cheese from the market, while another made coffee on the gas ring in the corner of the bench. One day there was talk about anaerobic bacteria on the bottom of a lake in Russia, or Leonardo da Vinci's engines of war or about poetry or printing. We never knew to what enchanted land we would be taken next."[44] Finances were a problem, though, and only a substantial gift of money from her aunt enabled her to survive.

During her first year at Cambridge, she was offered a post as tutor at Somerville College, Oxford. To entice her back to Oxford, the University agreed that she could spend a second year at Cambridge before taking up the position. Hodgkin accepted the offer and from then until retirement held appointments at Oxford. Once back in Oxford, she began an X-ray crystal analysis of cholesterol derivatives, but at every opportunity, she would return to visit the sparkling intellectual life at Cambridge, a contrast to the solitude of the Oxford environment. Her laboratory at Oxford was in the basement of the University Museum, completely separate from the chemistry facilities. Adding to the loneliness, women were not permitted to join the Oxford University Chemistry Club and although women could attend general sessions, they could not attend the weekly talks where current research was discussed.

In 1937, she acquired her first graduate student, Dennis Riley. Riley had invited Hodgkin to address the undergraduate chemistry club (at which women were welcome), and having become excited by her research, insisted that he wished to work with her. His request was not looked upon favorably by fellow chemists. Riley noted: "This, at the time, was quite revolutionary and several eyebrows were lifted. Here was I, a member of a prestigious college [Christ Church], choosing to do my fourth year's research in a new borderline subject with a young female who held no university appointment but only a fellowship in a women's college."[45] Riley enjoyed working with Hodgkin, but the conditions were far from pleasant as he commented:

> This [the X-ray room] was a peculiar room which always gave me a touch of claustrophobia. Owing to the architect's design of the University Museum in which we were housed, an ornate, unpractical Victorian structure, the room allotted to X-ray diffraction was a hybrid basement–ground-floor affair. The windows were high up so that from floor level one could not see out. There was a narrow balcony at window level reached by a steep ladder, and it was there that our polarizing microscope was placed and where we mounted crystals. Having done so, we had to descend the ladder [holding the ladder with one hand and the sample in the other], more or less precipitously, and fix our precious glass fibre or tube to the X-ray goniometer arcs.[46]

For Hodgkin, the maneuvering up and down the ladder was a particular problem, for during the early 1930s, Hodgkin's joints in her hands had become inflamed, tender, and painful. In 1934, the problem was diagnosed as rheumatoid arthritis and the disease plagued her for the rest of her life, only easing during her three pregnancies.

The same year (1937), Dorothy Hodgkin married the historian Thomas L. Hodgkin, and over the following years, she had three children. Thomas Hodgkin decided early in the marriage that Dorothy Hodgkin was the more creative of the two, and it was he who took the children to the dentist and the zoo and looked after them in the evenings while she returned to the laboratory. To add complication to their later lives, Thomas Hodgkin became active in African studies, which required them to have a residence in Ghana as well as England. Despite her work commitments, Dorothy Hodgkin always found time for her children, being able to switch between roles as mother and scientist with total ease. Fortunately, the social life improved, particularly through a friendship with Max Perutz, a scientist at Cambridge. Perutz would bring his wife and children to Oxford, where the two families would walk beside, and sometimes swim in, the river Thames (locally called "the Isis").[47]

Throughout her career, Hodgkin picked projects that were always just beyond the currently accepted limits of feasibility, and her research on the structure of cholesterol was one such example. This molecule had long been a puzzle to organic chemists, and it was Hodgkin and another of her students, Harry Carlisle, who determined the actual bonding arrangement, the first time that X-ray crystallography had been used to deduce an organic structure in which the atomic arrangement was unknown. This work was completed in the early years of the Second World War.[48]

In 1942, Hodgkin embarked upon the first groundbreaking study of her career—the molecular structure of penicillin. Structural studies were essential to help in the synthesis of penicillin, a task which was of vital importance during the war. Luckily, she had met Ernst Chain, the famous biochemist, some years earlier in the streets of Oxford, and he had promised to provide her with crystals of the antibiotic. Even to start the project was difficult, for it was not realized in those early days that penicillin could adopt different packing arrangements, depending upon the conditions of crystallization. In addition, no one knew the chemical organization of the molecule; in fact, much of the supposition of the time was proved to be incorrect. To help in the work, Hodgkin had one graduate student, Barbara Rogers Low. About the middle of the Second World War, they acquired nighttime use of an IBM punched card machine in an evacuated government building. Low helped write the first three-dimensional computer program and punched the data onto cards. The structure was finally completed in the summer of 1945, but of equal importance, Hodgkin had attacked the problem by a variety of different crystallographic techniques, broadening the options available for structure determination.

Her research group steadily expanded in the postwar era, and it included workers from around the world,[49] one researcher being Margaret Thatcher, who later changed her career direction to become British Prime Minister. Hodgkin decided to limit the numbers to about 10, to maintain the interactive environment of the group. A Rockefeller Foundation investigator commented that the lab was "...kept under good strong scientific discipline by their gentle lady boss who can outthink and outguess them on any score. A lovely small show reflecting clearly the quality of its director.... She conducts the affairs of her small laboratory on a most modest, almost self-deprecatory scale."[50] Though she had made important contributions to science, at the end of the war, her rank at Oxford was still only that of tutor. Deeply in debt, she realized that most of her male colleagues had university positions as well as research appointments so she asked Cyril Hinshelwood, Professor of Physical Chemistry, to help her acquire a better position. With his help, she was appointed as a university lecturer and demonstrator in 1946.

In 1948, a researcher with the Glaxo pharmaceutical company gave Hodgkin some deep red crystals of vitamin B_{12} that he had just obtained. It had been in 1926 that raw liver was shown to cure the disease of pernicious anemia, and the crucial compound had been identified as this vitamin. Before the vitamin could be synthesized, it was essential to determine the chemical structure. With 93 non-hydrogen atoms, many crystallographers regarded the task as impossible. Hodgkin disagreed. Over the following six years, the group grew more and larger crystals while they took a total of 2,500 X-ray photographs. The data thus acquired, the next task was the analysis, which was far beyond the normal calculating facilities of the time. A visit to Oxford by Kenneth Trueblood from the University of California at Los Angeles (UCLA) was to make this step possible. At UCLA, Trueblood had programmed one of the first high-speed electric computers for crystallographic calculations, and he had free computing time on the machine. He offered to help, and it was arranged that Hodgkin would send him the data, which he would run through the machine. He would then mail back the results.[51] In 1956, the structure of vitamin B_{12} was finally determined, and it was the largest chemical structure established up to that time. Once the structure was known, R. B. Woodward and Albert Eschenmoser were able to develop a synthetic route to vitamin B_{12}, a task which they completed in 1976.

Hodgkin was not promoted to university reader (approximately the equivalent of a North American full professor) until 1957, and even then she was not provided with modern lab facilities until the following year. The academic pinnacle of success, an endowed chair, was offered to her in 1960, but it was provided by the Royal Society, not Oxford University. Worldwide recognition of her work on the determination of the structures of biochemically important molecules came in 1964 with the Nobel Prize in chemistry. However, indicative of the attitude toward women scientists, the news was announced by one British newspaper as "Nobel Prize for British Wife". Some crystallographers felt that Hodgkin should have received the Nobel Prize earlier. For example, Max Perutz, who was co-recipient of the Nobel Prize for chemistry in 1962, commented:

> I felt embarrassed when I was awarded the Nobel Prize before Dorothy, whose great discoveries had been made with such fantastic skill and chemical insight and had preceded my own. The following summer I said as much to the Swedish crystallographer Gunner Hagg when I ran into him in a tram in Rome. He encouraged me to propose her, even though she had been proposed before. In fact, once there had been a news leak that she was about to receive the Nobel Prize, but it proved false; Dorothy never mentioned that disappointment to me until long after. Anyway, it was easy to make out a good case for her; Bragg [W. L.] and Kendrew [Perutz' Nobel co-recipient] signed it with me, and to my immense pleasure it produced the desired result soon after.[52]

Her third major project was to determine the structure of the protein, insulin. This was the culmination of 30 years of work, since her first X-ray photograph of the compound had been taken in 1935, and as early as 1939 she had published a report of X-ray measurements on wet insulin crystals. As techniques had improved over the years, she had kept returning to this particular molecule, but it was only technical advances in the 1960s that made the solution finally possible. When the results were published in 1969, the researchers were listed in alphabetical order, showing her willingness to share the credit and her egalitarian attitude toward all the research workers. In addition, she had one of her young postdoctoral fellows give the first lecture on the structure so that the glory was again shared rather than focused on herself.

Hodgkin continued the philosophy of the Braggs and of Bernal that science was a social activity. One of her later students, John H.

Robertson, recalled the family atmosphere of the Hodgkin group and particularly the afternoon tea:

> Each member of the community took his or her turn, weekly, to provide the little cakes that went with the afternoon's cup of tea. When anyone had a birthday, or a new baby, or anything comparable to celebrate, it was, by unwritten rule, that person's duty to provide a large iced cake, free, for that occasion. Each person had his or her own desk, of course, but everyone knew, at least in outline, what everyone else was doing. All the problems, and everything that was going on, were interesting. Mutual assistance was frequent; animated, even heated discussions were normal. The motivation was the interest of the subject. Everyone worked hard. It came naturally to do so.[53]

Like Lonsdale, Hodgkin had a very strong sense of social responsibility. Four of her mother's brothers had been killed in the First World War and her mother had been a strong supporter of the League of Nations. As a result of her beliefs, Hodgkin became a member of the 1930s Cambridge Scientists' Anti-War Group and, during the Second World War, chose to demonstrate her pacifist beliefs by the work on the penicillin structure, which she felt had the potential of saving lives.[54] After the War, she became a member of the Science for Peace organization. Membership in this organization caused her to be denied a visa to attend a meeting in the United States during 1953. For the next 27 years, to attend scientific meetings in the United States, she had to obtain a special entry permit from the U.S. Attorney General. Only in 1990, at the age of 80, did the State Department relent and approve a visa application.

It is difficult to overstate the challenges that Hodgkin faced in determining the structures of these very complex molecules. As one of her contemporaries, J. Dunitz, commented: "Dorothy had an unerring instinct for sensing the most significant structural problems in this field, she had the audacity to attack these problems when they seemed well-nigh insoluble, she had the perseverance to struggle onward where others would have given up, and she had the skill and imagination to solve these problems once the pieces of the puzzle began to take shape."[55] Hodgkin formally retired in 1977, but she continued to be active in science until her death on 30 July 1994.

Rosalind Franklin (1920–1958)

Whereas Hodgkin's life was one of success, Franklin died without recognition of her contributions to X-ray crystallography, and to science as a whole. Franklin was born in London, the second of five children of an affluent Jewish family.[56] Both her father, Ellis Franklin, and her mother, Muriel Waley, had family backgrounds as social activists. Ellis Franklin, a banker, taught science as a volunteer at working-men's clubs and helped Jews fleeing from Nazi Germany, and many of her aunts were practicing socialists and women's-rights workers. One of her uncles had even been jailed for his suffragette activities. It is from such a background that we can see why Franklin would not fit the passive, acquiescent role expected of women at that time.

Her mother concisely described those attributes of Franklin's character which were to lead to antagonisms later in life:

> Her affections both in childhood, and in later life, were deep and strong and lasting, but she could never be demonstrably affectionate or readily express her deeper feelings in words. This combination of strong feeling, sensibility and emotional reserve, often complicated by an intense concentration on the matter of the moment, whatever it might be, could provoke either a stony silence or a storm.[57]

During her early years, Franklin felt discriminated against because she was female, and as she recalled, her childhood was a battle for recognition. She detested dolls, much preferring sewing, carpentry, and Meccano construction sets. She was fortunate to be educated at St. Paul's Girls School, a strong academic institution that excelled in the teaching of physics and chemistry. Astronomy was her favorite hobby, and by the age of 15 she had decided to become a scientist. Franklin took and passed the entrance examinations for Cambridge University where she planned to study physical chemistry. However, her father refused to pay for her. He had once wanted to be a scientist, and he insisted that he would have been delighted for a son to follow such a path, but in his view, daughters should only consider volunteer work rather than a full-time career. Outraged at this attitude, both Franklin's favorite aunt and her mother offered to pay, and in the family crisis that followed, Ellis Franklin finally relented. As McGrayne comments, "his approval was grudgingly given and resentfully received."[58]

Franklin entered Cambridge University in 1938. With the start of the Second World War in 1939, the senior scientists were given war-related research and their time with students dwindled. With the lack of supervision, Franklin, working in the Cavendish Laboratory, enjoyed the opportunity for independent work. However, in the process, she seems to have driven herself to exhaustion, a possible cause for her obtaining an upper-second-class degree rather than a first-class one—another disappointment in her life.

After graduation in 1941, she accepted a position with the future Nobel laureate, Ronald Norrish, exploring gas-phase chromatography of organic mixtures. Franklin and Norrish had an abrasive relationship, which was due to personality problems on both sides. Franklin had been used to working very much on her own rather than having a supervisor telling her what to do, and this obviously made life difficult for both of them. Norrish also objected to Franklin's belief in sexual equality among scientists.[59] Franklin resigned from the position after one year in order to contribute to the war effort by doing research for the British Coal Utilization Research Association (BCURA). Her studies at BCURA contributed significantly to the understanding of coal structure and the effect of heating coals. Her classic research work helped found the science of carbon fibers, and it has proved important for the study of graphite nuclear reactor systems. She wrote up some of this work as her Ph.D. thesis, the degree being awarded in 1945.

During the war, she became friends with the French metallurgist, Adrienne R. Weill, a former worker with Marie Curie and Irène Joliot-Curie at the Institut Curie, and after the war, she wrote to Weill: "If ever you hear of anybody anxious for the services of a physical chemist who knows very little about physical chemistry, but quite a lot about the holes in coal, please let me know."[60] Weill suggested Marcel Mathieu, a former student of W. H. Bragg, [61] and Mathieu invited her to France in 1947 to work as a researcher in the Laboratoire Centrale des Services Chimique de l'État. While in Paris, she worked on the growth of graphite crystals with Mathieu's student, Jacques Méring, with whom she had a very positive relationship. Méring introduced her to the crystallographic techniques which were to become her field of expertise.

The time in France seemed to have been the happiest of her life. Franklin took science very seriously and in the laboratory she was intense and reserved, but outside she sparkled in the social life of Paris. Her co-workers were also friends with whom she had lunch at local bistros, went to dinner parties, and joined for skiing and mountain-climbing vacations. She particularly loved mountains and went on frequent long hikes and bicycle trips. Holidays, in fact, were one of her greatest joys, and they were planned in meticulous detail. As her mother remarked:

> Not for her the lazy holidays basking in the sun. Sunshine
> and warmth she loved, but her holidays must be full of

movement and never idle. Long walks of up to twenty miles a day; climbing, swimming; visits to picture galleries and ancient buildings, these were her delight. She was an eager stimulating companion, but those who travelled with her must possess a zest, an energy comparable to her own.[62]

After three years in Paris, Franklin accepted an offer in 1951 to work at King's College, University of London. Physicist John Randall, who had formed an interdisciplinary group of physicists, chemists, and biochemists, needed an expert crystallographer to analyze the X-ray photographs of DNA taken by Raymond Gosling, a graduate student. Having heard of Franklin's skills, he offered her the job, noting: "This means that as far as the experimental X-ray effort is concerned, there will be at the moment only yourself and Gosling, together with the temporary assistance of a graduate from Syracuse, Mrs. Heller."[63] When Franklin arrived, Randall's second-in-command, Maurice Wilkins, was away, and this proved to be a major problem, for Franklin assumed that she was working completely independently, while Wilkins, on his return, assumed that Franklin had been hired as a technical assistant to produce data for other members of the structure group to analyze. This was the start of the unhappy relationship between them.

Adding to the source of conflicts, the social life at King's was, as McGrayne contends, far different from Franklin's previous experience: "A number of women scientists worked on the staff, but they were not allowed to eat in the men's common room; women ate outside the lab or in the students' cafeteria. After work, the men visited a male-only bar for beer and shoptalk; women were not invited. As a result, the men talked science casually among friends while the women operated in a more formal office atmosphere."[64] This perspective is slightly different from the views of some of the other women who worked with Randall during that time period.[65] For example, Dame Honor Fell, another member of the group, considered that much of the social interaction took place in after-lunch coffee gatherings in one or other research lab. Franklin's "brusque manner" and her "overriding passion" for science meant, according to Fell, that Franklin never developed friendships, even with the eight women of the 31-member group. Whatever the cause, exclusion from social interaction not only meant a loss of companionship but also prevented her from understanding the hierarchical nature of Randall's group.

*Rosalind Franklin (1920–1958) mountain climbing in France, ca.1950.
(Courtesy of Jenifer M. Glynn.)*

Mainstream crystallography was then concerned with finding precise atomic locations in crystals of pure simple substances. However, the purpose of X-ray studies of complex materials, such as coal and DNA, was to gain a more general view of the atomic arrangements. Over the two years that she spent at King's, Franklin established that there were two forms of DNA, A and B, the form depending upon the humidity, a point that had confused the other researchers in the field. Franklin developed techniques to produce the most high-resolution X-ray photographs of DNA taken up to that time. She showed that the phosphate groups were on the outside of the DNA molecule and that hydrogen bonding played an important role.

Gosling enjoyed working with Franklin and he commented on her strong personality and ebullient, argumentative character. These attributes did not endear her to Wilkins as Wilkins himself was "shy, passive, indirect."[66] As a result, the relationship between the two, which had never been positive, deteriorated into active dislike.

During this period, James Watson and Francis Crick were constructing their models of DNA at Cambridge University. Wilkins developed a friendship with the two, confiding in them about his difficulties with Franklin. Without telling Franklin, Wilkins removed the superb X-ray photograph of the B-form and showed it to Watson and Crick, providing the major experimental evidence for the helical structure of DNA. Franklin had summarized her work in a government agency report and this had been distributed to the agency review committee. A member of this review committee passed the report, unethically in the view of many scientists, to Watson and Crick, providing additional essential information.

When Watson and Crick submitted their famous article on DNA structure to the journal *Nature*, the editor contacted both Wilkins and Franklin to ask if they would submit accompanying articles. Franklin had by this time herself deduced the helical nature of DNA but she had not progressed as far as Watson and Crick. Whether she would have done so in time, or whether Watson and Crick could have solved the structure without Franklin's photograph and report, is still a matter of debate. The three papers appeared sequentially, with the vague comment by Crick and Watson that they had "been stimulated by a knowledge of the general nature of the unpublished experimental results and ideas of Dr. M. Wilkins, Dr. R. E. Franklin, and their co-workers at King's College, London."[67] Franklin was never aware that they had actually seen her photograph and report.

The protein crystallographer David Harker identified the key issue:

And the real tragedy in this affair is the very shady behavior by a number of people, as well as a number of unfortunate accidents, which have resulted in the transfer of
information in an irregular way.... I would never have consciously become involved in anything like this behavior,
especially the transfer of information through a privileged
manuscript. And I think these people are—to the extent
that they did these things—outside scientific morals as I
know them.[68]

By 1953, Franklin felt that relations with Wilkins had deteriorated to
the point that she could no longer work at King's. She was asked by
Kathleen Lonsdale to join her at University College[69], but instead
Franklin asked if she could join Bernal's group, then at Birkbeck College[70], the graduate night school of the University of London. Randall agreed to the transfer of her fellowship to Bernal, on condition
that she did not continue working on DNA (at that time it was not
uncommon in Britain for specific research fields to be the "property"
of particular research groups). Bernal arranged for Franklin to be the
head of her own research group, and he suggested that she look at
the structure of the tobacco mosaic virus, a topic that he had started
and abandoned years earlier. By working with Bernal and Mathieu,
Franklin was now doubly scientifically related to the Braggs.

In her five years at Birkbeck, she and her research group determined the structural features of the virus, showing it to be hollow-
cored, not solid as supposed by microbiologists, and she was able to
determine surface features of the virus for the first time. During this
period, her group was the world leader in the structure of viruses, and
for the first time she was able to develop positive collaborations with
other research groups. Of equal or greater importance, she was back
in a pleasant working environment. However, there were still
moments of frustration, such as the refusal (at the time) of the British
Agricultural Research Council to fund any project that had a woman
directing it. Fortunately, the U.S. Public Health Service provided her
with adequate funding instead.

During 1956, she experienced extreme pain which was diagnosed
as ovarian cancer. Three operations and experimental chemotherapy
followed, but without success. During these last years, Franklin had
become friends with Francis Crick and his French spouse, Odile,
showing no animosity for the DNA dispute. While convalescing with
them, Francis Crick had suggested that she move her group to Cambridge and this she decided to do. Realizing that she was terminally
ill, Franklin started a dangerous individual project, the structure of

the live polio virus. This work was so hazardous, that after her death, the project was halted. She died on 16 April 1958 at the age of 37.

Four years later, the Nobel Prize was awarded to Crick, Watson, and Wilkins, for the DNA work. The Nobel Prize is only awarded to living scientists and to no more than three people in any category. Having died, she was no longer eligible, but the controversy focused upon the lack of acknowledgement of her work during the Nobel lectures by the winners.

Many scientists argued that the popular account of the discovery of DNA written by Watson[71] minimized Franklin's contributions and painted a very negative picture of her personality. As her biographer, Sayre, has commented, everyone is entitled to their own perception of another individual, but the picture of Franklin was to generate a totally misleading image which would reinforce the negative stereotype of women scientists. Sayre herself remarked:

> "Rosy" [Watson's nickname for Franklin] was less an individual than...a character in a work of fiction.... If Rosalind was concealed, the figure which emerged was plain enough. She was one we have all met before, not often in the flesh, but constantly in a certain kind of social mythology. She was the perfect, unadulterated stereotype of the unattractive, dowdy, rigid, aggressive, overbearing, steely, "unfeminine" bluestocking, the female grotesque we have all been taught to fear or to despise.[72]

In fact, her closest collaborator at Birkbeck, Aaron Klug, was so incensed at the belittling of her role that he had a strong attack on her detractors published in the prestigious journal *Nature*.[73] A former colleague of Franklin's, A. J. Caraffi, wrote to thank Klug for his defense of Franklin:

> I would like to say how very much I have appreciated your paper to *Nature* on Rosalind Franklin and DNA. After the journalistic distortions to which her rightful place in the history of science has been subjected, it is most satisfying to one's sense of justice to see published a dignified, objective, and properly documented account of her signal contribution. It brings out her absolute integrity and illustrates admirably the thorough principles applied by great experimentalists to the evaluation of their proofs. There could be no better counter to the titillations of all the offensive "Rosie" nonsense that was bound to attract selection and emphasis by ignorant hacks [journalists].[74]

Bernal wrote a very positive obituary on Franklin, concluding with the comments: "As a scientist Miss Franklin was distinguished by extreme clarity and perfection in everything she undertook.... She did nearly all the work with her own hands. At the same time, she proved to be an admirable director of a research team and inspired those who worked with her to reach the same high standard."[75] Klug commented in a letter to P. Siekevitz of the New York Academy of Sciences:

> However, if she is to be honoured, it should not be as a "woman in science" but for her crucial contributions in sorting out the A and B forms, establishing that the phosphates were on the outside and determining the helical parameters which were used by Crick and Watson in their study.... There is also, inevitably, a fair amount of discussion as to whether she would have solved the structure on her own. One can only guess, but my view, as stated, is that she would have done so eventually....[76]

Women as Crystallographers

Unusual for their time, both Lonsdale and Hodgkin managed to raise a family as well as devote long hours in the laboratory. This was, to a large extent, a result of very progressive marriages, with the husbands performing some of the "wifely" duties. Franklin followed the more common single pattern. Though she expressed a preference for marriage, not wishing to remain a "spinster professor," she felt that children would have interfered with the research to which she wanted to devote her life. Hilary Rose, a sociologist of science, has suggested that the "single and sexually threatening" image contributed to Franklin's problems as compared with the safe and respectable perception of Hodgkin and Lonsdale as wives and mothers.[77] However, it can also be argued that Franklin's abrasive manner did not assist with her acceptance into British scientific society.

Crystallography seems to provide one of the clearest examples of the positive and supportive role of supervisors. And this effect continued until more recent times, for many of the Braggs' protégés, when they gained academic rank, accepted women students, such as Isabella Lugoski Karle (1921–).[78] Julian has shown that women have maintained a significant, though not overwhelming, presence in crystallography.[79] In 1981, women represented worldwide about 14% of the total, though the average did not show the uneven distribution;

for example, Thailand had the highest proportion of women crystal-lographers—14 of 28. One of Hodgkin's students, Judith Howard, has speculated that "...one of the reasons why women continue to be attracted to crystallography is because so many early women crystal-lographers were so good."[80] She added that, in 1995, about one-third of the British Crystallographic Association's 800 members were women.

Chapter 6

Women in Radioactivity

Early Research in Radioactivity—The Pioneer Women—The Mentors—Biographies: Marie Sklodowska Curie (1867–1934); Ellen Gleditsch (1879–1968); Catherine Chamié (1888–1950); Irène Joliot-Curie (1897–1956); Marguerite Perey (1909–1975); Harriet Brooks (1876–1933); Stefanie Horovitz (1887–1940); Lise Meitner (1878–1968); Ida Tacke Noddack (1896–1978); Maria Goeppert-Mayer (1906–1972)—The Change in the Nature of Atomic Research

Early Research in Radioactivity

The discovery of X-rays in 1895 led to the search for other types of radiation. Scientists claimed to find rays everywhere, and many years were to pass before valid claims were distinguished from the spurious reports, such as N-rays.[1] The first evidence of spontaneous radiation from a chemical compound was the darkening of photographic plates by uranium salts. This phenomenon was first observed by Sylvanus P. Thompson in England, but the French chemist Henri Becquerel gained the most recognition because, though his discovery of the rays was after Thompson, he published his findings first.[2] Nevertheless, the so-called hyperphosphorescence, or Becquerel rays, from uranium aroused little interest at the time. The rise in interest in 1898 resulted from the simultaneous discovery by Marie Sklodowska Curie in France and Gerhardt Carl Schmidt in Germany that thorium also emitted Becquerel rays.[3] And so the search was on for the nature of the rays and for evidence of radioactivity among other elements.

From this point on, the study of radioactivity and the radioactive elements grew at an astonishingly rapid rate[4], and by the end of 1907 more than 1,200 research papers had been published in the field.[5] Like most other aspects of science, many individuals were involved. At the beginning of the twentieth century there were three research foci—Marie Sklodowska Curie and Pierre Curie in Paris (the French school); Ernest Rutherford in Montreal, and later, Manchester, then Cambridge (the British school); and Stefan Meyer in Vienna (the Austro-German school).[6]

Each of the schools approached the study from a different direction. For example, the Curies, in the early years, emphasized the chemical aspects, such as the isolation of the radioactive elements; Meyer, on the other hand, was the pure physicist, interested in the nature of the radiations; and the Rutherford school bridged the chemical and physical aspects. These differences in philosophy determined which group had priority in discovery. For example, Rutherford and his co-researcher, Frederick Soddy, were the first to propose the theory of the transmutation of chemical elements even though Pierre Curie and his co-researcher, André Debierne, had the results from similar experiments at that time.[7]

The Pioneer Women

In addition to crystallography, the study of radioactivity was another field in which women scientists clustered.[8] Like crystallography, radioactivity was on the edge of mainstream science. In fact, radioactivity was a sort of "orphan" between chemistry and physics. As an example, Ernest Rutherford, who considered himself a physicist, was—to his surprise—awarded a Nobel Prize in Chemistry.

One of the most interesting aspects of the early women researchers in radioactivity was the interaction among them, which provided a valuable source of support. Such professional interactions are referred to as invisible colleges[9] for they link people who do not necessarily have affiliations to the same institution. For example, Ellen Gleditsch of the Paris group corresponded with and visited Lise Meitner of the Vienna/Berlin group.[10] Also, many of the researchers moved among the research centers. The travels of Elizabeth Róna,[11] who became the leading expert in polonium, illustrate this more than any other individual. During her hectic life, Róna worked with Meitner and Otto Hahn at the Kaiser Wilhelm Institute, Berlin; Meyer at the Radium Institute, Vienna; Curie and her daughter, Irène Joliot-Curie, at the Institut de Radium, Paris; and Gleditsch at the University of Oslo, Norway. She turned down an offer to join Rutherford at

Cambridge, England. All this occurred before Róna, of Jewish origin, fled from Nazi-occupied Europe to the safety of the United States where she later obtained a position at Oak Ridge Associated Universities, Tennessee.

It is clear that the women researchers regarded science as the purpose of their lives. For example, in a letter to Curie, the Swedish scientist Eva Ramstedt commented: "As for me, I have been unfaithful to science, never finding time for research."[12] In fact, women who excelled academically were expected to consider it an honor and a duty to devote their lives to knowledge. This parallel to a religious vocation, excluding the possibility of family life, is illustrated by the following contemporary quote: "Civilization rests upon dedicated [women's] lives, lives which acknowledge obligation not to themselves or to other single persons, but to the community, to science, to art, to the cause."[13] Yet there was no equivalent requirement for male researchers to remain single (and most, in fact, were married).

The Mentors

These three groups—French, British, and Austro-German—attracted numerous research workers, many of them women. Those that ventured to Paris saw the famous "Madame Curie" as a role model. At least one of those individuals, Ellen Gleditsch, specifically dreamed of working with Curie.[14] However, Curie did not seem to act in any mentoring role. May Sybil Leslie, a researcher with Curie, commented: "She [Curie] does not appear to come around much to the students but receives them very kindly when they seek her".[15] As Curie's biographer Rosalynd Pflaum remarked: "And nothing she did was consciously done to open the doors for women who would follow her example."[16] Of particular relevance, Harriet Brooks[17] and Jadwiga Szmidt[18] worked with both Curie and Rutherford, but it was to Rutherford that they turned for support and encouragement.

Yet we must consider Curie in the context of her position and her time. The discoveries of Marie Curie automatically propelled her to fame, without question of her gender. And in those early years she was the one and only woman among the leading researchers in radioactivity. The historian Allison Heisch describes the problem of the exceptional woman: "Exceptional women are not representative women, and for many such women one condition of being both exceptional and female may be that the values and practice of the male society in which they function may be accepted by them, transformed and internalized, and followed, so that they become, in effect, 'honorary males.'"[19] Thus Curie identified with her fellow famous (male) scientists more than she did with the challenges facing the "normal" woman scientist.

Rutherford started his research career at the Cavendish Laboratory, Cambridge. This laboratory had been opened to women in 1880 at the insistence of Lord Rayleigh, at a time when other laboratories specifically excluded women.[20] Rutherford made his views on women in science quite explicit in a letter to *The Times* of London. The cause of the letter was a furor at Cambridge University over whether women should be admitted to the university with the same privileges as men. Rutherford and his coauthor, William Pope, wrote: "For our part, we welcome the presence of women in our laboratories on the ground that residence in this University is intended to fit the rising generation to take its proper place in the outside world, where, to an ever increasing extent, men and women are being called upon to work harmoniously side by side in every department of human affairs."[21] Of equal importance, Rutherford was an active mentor to his young research workers.[22] As one of Rutherford's later collaborators, Peter Kapitza, noted, "He [Rutherford] was also very particular not to give a beginner technically difficult research work. He reckoned that, even if a man was able, he needed some success to begin with. Otherwise he might be disappointed in his abilities, which could be disastrous for his future. Any success of a young researcher must be duly appreciated and must be duly acknowledged."[23]

Little has been recorded of the workings of the Vienna group, led by Stefan Meyer. Austria was the major source of radioactive ores in the early years, and hence the Vienna Institute had a greater access to radioactive materials than any other research institute. Like Paris, about one-third of the research workers were women. The disintegration of the Austro-Hungarian empire after the First World War seems to have diminished the influence of the Vienna school and many of the scientists moved elsewhere, particularly to Germany where the Kaiser Wilhelm Institute at Berlin-Dahlem became a new focus. Included in this "brain drain" from Vienna was Otto Frisch, a later collaborator with Meitner.[24] Those that stayed in Vienna, like Marietta Blau,[25] tended to be pure physicists, studying such phenomena as cosmic rays.

It is probably the personalities of Rutherford and Meyer that had a major influence on the women who worked with them. In the case of Rutherford, his biographer, David Wilson, has commented: "It will inevitably be hidden and forgotten that he was a man of exceptional personal kindness. Everyone who remembers Rutherford remembers this—that he was personally kind to them far and away beyond the normal behaviour of a pleasant human being."[26] Meyer was regarded similarly. In an obituary, the prominent atomic physicist Fritz Paneth noted: "All those who at any time have worked in Meyer's Institute, as his assistants or as guests, remember with deep gratitude his never-failing kindness"[27]; and a former researcher with Meyer, Robert Lawson, commented on Meyer's "...personal charm and good nature, his warm friendship and his innate kindliness."[28] Róna, who worked at Vienna for many years, remarked: "The atmosphere at the institute was most pleasant. We were all members of one family. Each took an interest in the research of others, offering to help in the experiments and ready to exchange ideas. Friendships developed that have lasted to the present day. The personality of Meyer and that of the associate director, Karl Przibram, had much to do with creating that pleasant atmosphere."[29]

As we have already discussed, it is reasonable to argue that there were three founding schools of radioactivity, one of which was led by Marie Curie. Thus we will first consider Curie's life and contribution, and then continue with the women of the second generation—those that were born at the close of the nineteenth and the beginning of the twentieth century.

Biographies

Marie Sklodowska Curie (1867–1934)

Of all women scientists, "Mme. Curie" is by far the best known, and numerous biographies of her have been published.[30] Yet we have to be careful to provide a balanced portrait, for several of these biographies are hagiographies, presenting a saintlike image of "our lady of radium", as the science historian Lawrence Badash has noted.[31] In addition, many accounts have stressed Curie as a "loner" with a slavish devotion to work, called the "Marie Curie Syndrome" by Stephen Brush,[32] rather than recognizing that there were up to 30 researchers in her later group and that she balanced research with her family commitments.

Curie, born Marya Sklodowska in Warsaw, Poland, was the fifth and youngest child of Wladyslaw Sklodowski and Bronislawa Boguska. Her father taught physics and mathematics while her mother was a full-time director of a private school for girls, a position she held until after the birth of Curie. During that period, Poland was divided between Germany and Russia, the Sklodowski family living in the Russian-controlled part. The family held fervent Polish nationalist views, which resulted in her father being fired from job after job. As a result, Curie grew up in an increasingly impoverished environment. Her mother, a sufferer of tuberculosis, never held or kissed Marie for fear of transmitting the disease. Curie's oldest sister died of typhus in 1876 and her mother of tuberculosis in 1878. These two tragedies drove Curie into depression and to embrace atheistic views. Thus the later image of Curie was formed largely in her youth: the expectations of combining work and family; her Polish nationalism; the disinterest in material possessions; the unemotional exterior; and the rejection of religion.

Curie excelled at school, even though the workload and the constant stress of suppressing her nationalistic views was to result in the first of several physical breakdowns. At the age of 15, she graduated high school, first in all subjects. She wanted to continue her studies, but at the time, the Russian government forbade women from attending university. Thus she made a pact with her older sister, Bronislawa (Bronya), that Curie would help support Bronya through medical school in Paris, then Bronya, in turn, would help pay Curie's way through university. Curie took a governess position—more accurately, a servant role—with the Zorawskis, a wealthy Polish family. While there, she developed a passionate relationship with the eldest son, but the Zorawskis refused them permission to marry because of her low status in Polish society. After her contract with the Zorawskis was completed, Bronya, who played an influential and supportive role in Curie's life, tried to persuade her to move to Paris but initially without success. Instead, Curie moved in with her father and joined Warsaw's "underground" Polish university. There she was given laboratory space where she practiced experiments from physics and chemistry textbooks and where she first acquired her fascination for experimental research.

In 1891, with her taste for research developed, her hopes for marriage to the younger Zorawski finally abandoned, and enough money to support herself, Curie finally traveled to France. At this time, she changed her name from Marya to the French form, Marie. Though Bronya provided Marie Curie with a room in the small apartment where she, her husband, and baby lived, Curie soon moved out, preferring seclusion and privacy. Her grasp of the French language was initially quite weak; nevertheless, Curie excelled academically at the University of Paris, earning the equivalent of a master's degree in physics in 1893, placing first in her class, and a similar degree in mathematics the following year. In 1894, she met Pierre Curie, a professor at the Municipal School of Industrial Physics and Chemistry. The relationship developed rapidly, but Marie Curie rejected Pierre's proposals of marriage or cohabitation as she intended to return to Poland to teach. Only when Pierre offered to give up his research career *and* move to Poland did she consent to marriage. The civil ceremony took place in 1895, and they decided to stay in Paris while Marie completed a teaching certificate and Pierre finished his doctorate. Professor Schutzenberger of the Municipal School arranged that the Curies could work together in a laboratory there. It was the marriage that gave Marie the opportunity of a career in scientific research. The historian Helena Pycior has commented:

> Had Marie Curie not married Pierre in 1895, she would quite possibly have had no scientific career of note. The marriage not only rescued Curie from a perhaps obscure career as a teacher in her native Poland, but also assured her an understanding, scientific husband, who, first of all, found her laboratory space and, secondly, served as an essential link with the French male scientific community.[33]

Two years after their marriage, a daughter, Irène, was born, and Pierre's widowed father, Eugène, moved into the household to look after the baby. Marie's first research work was the study of the magnetism of steel, a paper that itself gained international respect.[34] This, like many of her early papers, was published under the name of Mme. Sklodowska Curie, no doubt to give recognition to "Polish science." Curie then decided to embark on a doctoral thesis.[35] Though Becquerel's discovery of rays from uranium had aroused little interest among the scientific community, she chose as her topic the search for rays from other elements. This one decision initiated her path to fame. Of particular importance, she chose to do the search using a sensitive gold-leaf electroscope rather than the slow and less responsive method of the darkening of photographic film. Within days, she had discovered that thorium, too, released rays, and she proposed the term *radioactivity* to describe the phenomenon. The radioactivity of thorium was reported at the same time by Schmidt in Germany,[36] but the next step was the crucial one. Marie tested museum samples of the uranium ores, pitchblende and chalcolite, for radioactivity and found that they were several times more radioactive than either uranium or thorium. In a paper in April 1898,[37] she hypothesized that the ores contained a hitherto unknown highly radioactive element. At this point, Pierre ceased his own research on piezoelectricity and joined Marie in an attempt to isolate this new element. During this time, Marie taught part-time at the women teachers' college at Sèvres. By the middle of the year, the pair had extracted a bismuth-containing precipitate that was highly radioactive. This precipitate, they concluded, must contain traces of a previously unknown radioactive element. Marie named this element polonium, after her native country.[38]

While extracting the polonium, the Curies and Gustave Bémont obtained a precipitate of a barium salt that was extremely radioactive. They argued that the precipitate contained, as impurity, a radioactive barium-like element.[39] This element was obviously a very minor component of pitchblende; hence a large quantity of the ore—several tons—was necessary for the extraction of measurable quantities of this next new element. Pitchblende was prohibitively expensive, but the Curies were able to obtain several tons of insoluble waste material from the pitchblende mine at St. Joachimsthal in Bohemia (now part of the Czech Republic). The first extraction steps were performed by the Curies' friend and collaborator, André Debierne, at the Société Centrale de Produits Chemiques. This was the time when the legend of the Curies working in a shed was born, for the only place large enough for the subsequent extraction steps was the abandoned dissection shed of the Municipal School. The visiting scientist Wilhelm Ostwald remarked: "It was a cross between a horse-stable and a pota-to-cellar and if I had not seen the worktable with the chemical apparatus, I would have considered it a practical joke."[40] Apart from the heat in summer and the cold in winter, the work was arduous, involving the stirring of boiling solutions and slurries containing up to 20 kg of material in large cast-iron basins. Their efforts were successful and by the end of 1898, they announced the discovery of this highly radioactive element that they named radium.[41] The light emitted by this element fascinated them; Marie kept a sample at her bedside and Pierre carried one in his pocket. Marie remarked: "One of our joys was to go into the workroom at night; we then perceived on all sides the feebly luminous silhouettes of the bottles or capsules containing our products. It was a really lovely sight and always new to us. The glowing tubes looked like faint fairy lights."[42]

In modern discussions of the Curies, the question is sometimes raised as to the role of each.[43] The initial discovery of the radioactivity of thorium had been made by Marie alone. The isolation of the radioactive bismuth and barium precipitates had been joint Curie projects, the latter also involving Bémont. Though both Curies had physics backgrounds, it would seem that at this point, Pierre became more interested in the phenomenon of radioactivity while it was Marie who veered into chemistry. For example, Marie reminisced, "Sometimes I had to spend a whole day mixing a boiling mass with a heavy iron rod nearly as large as myself. I would be broken with fatigue at the end of the day."[44] Though the Curies continued to publish joint papers on the physics of radioactivity, the chemical ones were authored by Marie. Marie had probably learned the chemical precipitation techniques from Debierne as he was the only one of the group who had a strong chemical background. Debierne was given a simple acknowledgment in the paper on the discovery of radium but he was sole author of the paper on the discovery of actinium,[45] another element extracted from the pitchblende residues.

Marie had earlier received an award from the French Academy of Sciences for the polonium discovery. Now the possibility of a Nobel Prize arose. In 1903, the same academy submitted the names of Henri Becquerel and Pierre Curie as candidates for the Nobel Prize in Physics for their work on radioactivity. The omission of Marie's name from the recommendation was noticed by the Swedish scientist, Magnus Gösta Mittag-Leffler. Mittag-Leffler had been a strong supporter of women scientists, and he had, against tremendous opposition, obtained a faculty position at the University of Stockholm for the renowned Russian mathematician, Sofia Kovalevskaia[46] (*see* discussion of Yulya Lermontova in Chapter 4). Though Marie Curie had not been nominated for 1903, Mittag-Leffler recalled that she had been nominated in 1902, and he and others proposed that one of the 1902 nominations was still valid for the 1903 competition. By this means, Marie could be considered along with Pierre. Becquerel and the Curies did indeed receive the Physics Prize for that year, but the chemists on the committee insisted that the language make clear that the prize had been awarded for the work on radioactivity, not for the discovery of radium. The chemists argued that the isolation of radium was of sufficient importance in itself for a subsequent nomination for a Nobel Prize in Chemistry. Both Curies were too ill to collect their awards in 1903, and it was another eighteen months before Pierre was able to travel to Stockholm to give the required lecture and collect the prize money.

Though fame immediately arrived, including offers of professorships at the University of Geneva, Pierre did not receive a professorship at the Sorbonne of the University of Paris until 1904. Marie was promoted from lecturer to professor at the women teachers' college at Sèvres, her dreams of returning to Poland forgotten. She was told that she would become superintendent of Pierre's laboratory when it was built at the Sorbonne, but many years were to pass before con-struction started. To add to their joy, in 1904, their second child, Ève Curie, was born.

Everything changed on 19 April 1906. Pierre Curie was killed in a street accident. After some hesitation, the University of Paris offered Marie an assistant lecturer position, effective 1 May 1906, making her the first woman professor in the 650-year history of the University. Thus her academic post was only secured as a result of Pierre's death. It is questionable how much longer Pierre would have lived in any event, for even in 1903, he had been showing signs of severe radiation poisoning, in part from his laboratory exposure and in part from car-rying the radium sample in his pocket at all times.

The next challenge to face Marie was the announcement by Lord Kelvin that radium was not a new element but simply a compound of lead and helium, a possibility that would account for the release of α-rays (helium nuclei) as radiation. The announcement of the new element had been based on the production of radium chloride, rather than the element itself, and it took four years of effort for Marie Curie and André Debierne to isolate pure radium metal by electrolyzing molten radium chloride and prove Lord Kelvin wrong.

The year 1911 started disastrously for Marie. First, she announced her candidacy for election to the French Academy of Sciences. Until her announcement, it was assumed that Edouard Branly, who was the first to show that radio transmissions could be performed without a wire, would be chosen to fill the vacancy. Branly was elderly, French, and a devout Catholic, and the competition became a public event, pitting the liberal, feminist, and anticlerical supporters of Curie against the nationalist, antifeminist, and Catholic supporters of Branly. Curie lost by one vote. But the worse was yet to come. The Curies had been friends with the young physicist, Paul Langevin, and after Pierre's death, the friendship between Langevin and Marie con-tinued and deepened, probably to that of a physical relationship.[47] Jeanne Langevin, his spouse, filed for separation on the basis of adul-tery, and the right-wing Paris newspapers used the issue to point to Curie's foreign nationality. This might well have been the end of Marie's career, but for the announcement that she had been awarded the Nobel Prize in Chemistry for her discovery of radium. The timing may not have been coincidental. There had been only two nomina-tions of Curie, one of them by Svante Arrhenius, a powerful figure on the Nobel selection committee, and himself a strong proponent of women scientists. Thus it is quite possible that he pushed through her

Marie Curie (1867–1934) in her laboratory, ca. 1910. (Courtesy of Culver Pictures.)

nomination to restore her scientific reputation, for many at the time argued that Pierre, not Marie, had been the truly creative scientist.[48]

During this period of 1910–1912, Curie had one great success. In the spring of 1910, Rutherford had been comparing the radioactivity of his own radium samples with one from Meyer in Vienna, and he requested a sample from Curie for comparison. In the Rutherford–Curie correspondence, they discussed the possibility of forming an international committee to establish a standard of radioactivity. Following this exchange, at the International Congress of Radiology and Electricity held in 1910, it was proposed that Curie would be responsible for preparing the standard sample. A year later, the leaders of the different radioactivity research groups met in Brussels for the first Solvay Conference, named after Ernest Solvay, a chemical industrialist, who sponsored the meeting. There were a number of conflicts. For example, Curie initially proposed that there was to be only one standard and that it had to be kept in her laboratory. Meyer, on the other hand, wanted his sample to be compared with that of Curie's and to be declared a standard as well. Curie was not in the best of health, which did not help, as her biographer Rosalynd Pflaum has commented: "Marie complained frequently of headaches and these, combined with her automatic attacks of nervousness whenever she had to address an audience, caused her frequent, abrupt departures from platforms—and even from committees—often at crucial moments."[49] Throughout the meeting, Rutherford acted as a mediator, always sympathetic to her, but firm that the standard had to be in impartial hands. In this he succeeded, at the same time supporting her demand that the unit of radioactivity be called a curie. The following year, the Paris and Vienna samples were compared and their radioactivity found to be in very close agreement. Subsequently, Curie and Debierne deposited their standard sample with the Office of Weights and Measures at Sèvres, where the standard meter and liter were held, establishing the curie as a true standard of international measurement.

Unfortunately, Curie's problems—the Langevin affair, her own recurrent illnesses, and the responsibility of two daughters—coincided with a period in radioactivity research when the greatest number of advances were being made. Susan Quinn, another biographer of Curie, has noted:

> And yet, despite her keen interest and understanding of the new developments in physics, Marie Curie's contribution after 1910...came mainly through the work of others in

her laboratory. Her own efforts would continue largely in the byways of radiochemistry. And while it is impossible to know whether she would have broken more new ground without the interruption of the Langevin scandal and her illness, her personal difficulties certainly affected her productivity. More important, all the hurt and humiliation accentuated her tendency to defensiveness. Much of the energy she had, during these critical years in radioactive research, was used up in defending her past accomplishments.[50]

When the First World War started, all of the male researchers in the laboratory were drafted. Marie stopped her research work and began organizing a mobile X-ray service for hospital units in the front lines. There were X-ray facilities in the major hospitals, but Curie argued that X-ray services needed to be provided at the battle front. Against opposition from all levels in the medical and military organizations, she found allies to allow her to organize the vehicles and equipment. Both she and Irène served as X-ray technicians until 1916, when the government, having realized the importance of X-ray technology, asked them to teach classes of women in the subject; the graduates of these classes were sent immediately to front-line posts. By the end of the war, these X-ray units had examined more than one million soldiers.

Following the war, the Institut de Radium was built, an establishment where at last Marie and her band of research workers could have reasonable working space. The administration duties became a major burden, again affecting her own research output. In 1931, for example, there were 37 researchers under her guidance, 12 of them women. Opinions of "la patronne" differed; some found her to be kind and generous, while others, particularly in later years, found her to be cold and dictatorial. One contentious issue was her advancement of Irène Curie as her heir and successor over other talented researchers in the group. Certainly the Institut was a unique environment, described as having an atmosphere like a religious convent, containing people who had dedicated their lives to radioactivity.

Though research into the chemistry and physics of radioactivity continued at the Institut de Radium, Curie had always hoped to find medical uses for the radiation. As early as 1915, claims had been made for radium therapy—called "curietherapy" in France—as a means of treating a wide range of diseases including various types of cancer. Curietherapy could take such diverse forms as ingesting solutions of radium salts, inhaling radon gas, being washed in dilute radium solutions, or having injections of radium solutions. Cancers were usually treated by placing an ampoule of the radiation source close to the tumor. Curie held to a belief in the beneficial properties of curietherapy, and the possible medical value of radium became her focus in life.

In 1920, she was visited by an American journalist, Marie Mattingly (Missy) Meloney. Hearing that Curie had less than one gram of radium for her research, while other laboratories around the world had much more, Meloney vowed to raise funds to augment the Institut's supply. She arranged a three-week tour of the United States for Marie, Irène, and Ève to promote the fundraising effort. The Curie family arrived to newspaper headlines that radium was going to provide cures for all cancers. This hyperbole, much more effective than any mention of the importance of fundamental research, enabled Meloney to turn the visit into a publicist's dream. To aid her cause, Meloney portrayed Curie as an impoverished, poorly funded scientist. This depiction had been true 10 years earlier, but Curie now possessed a laboratory that was the envy of other researchers and her income had become very respectable. As a result of the poverty-stricken image of Curie, which largely survives even today, contributions flooded in. Marie was awarded honorary degrees from 20 American universities and even had a reception with President Harding. Yet despite this success, there was no specific encouragement of women scientists by Curie. As Sharon McGrayne has commented: "In fact, by creating an almost impossible standard for women scientists to live up to, Marie Curie may have made their professional progress more difficult. Although universities did not expect every male scientist to be an Albert Einstein, women scientists were continually measured against Marie Curie—and naturally found wanting."[51] Though the traveling and the public appearances exhausted Marie, a second successful tour took place in 1929.

Between these two trans-Atlantic expeditions, she underwent four cataract operations, cataracts being one of the first symptoms of radiation sickness. Curie refused to publicly admit that radiation was dangerous even after Sonia Cotelle, a Polish researcher in the Curie laboratory, died agonizingly after ingesting a polonium solution that she was pipetting. There were an abnormally high number of premature deaths among the Curie workers, and most would now be classified as radiation-related.[52] In 1931, 7 of 20 workers had blood abnormalities for which Curie prescribed country rest and lots of fresh air, believing any radiation-related illness to be a temporary condition. Yet, amazingly, Curie survived to the age of 67, dying of leukemia in

1934 at a sanitarium in the French Alps. Before her death, she destroyed many of her personal papers including all of those from the Langevin period; hence we will never gain a full understanding of her complex personality.

Ellen Gleditsch (1879–1968)

As we mentioned earlier, the popular image of Marie Curie is that of the lone toiler in the unheated shed, separating radium from pitchblende residues. Yet this was but a short period in Curie's professional career. Following her first Nobel Prize, requests to work with her exceeded the space available in the Curie laboratory. Curie's sole mention of the legions of research workers at the Institut de Radium was the comment in her autobiographical notes concerning the continuation of the research after Pierre's death: "A few scientists and students had already been admitted to work there with my husband and me. With their help, I was able to continue the course of research with good success."[53] Any mention of the contributions of others was perceived as detracting from the Curie name and reinforcing the critics' view that Pierre Curie had really provided the genius behind their work. The individual who suffered the most from living in the shadow of the Curies was the faithful friend and colleague, André Debierne.[54] Unfortunately, there are few records of the visiting researchers—only their names and publications appear in the Annual Report of the Institut de Radium. One individual about whom we do have a significant quantity of information is the Norwegian chemist Ellen Gleditsch[55], and her life was interwoven with that of the Curie family.

Gleditsch was born in 1879 in Mandal, a small town in southern Norway.[56] She was the eldest of 10 children; her father, Karl Kristian Gleditsch, was a teacher, and her mother, Petra Birgitte Hansen, the daughter of a sea captain. Hansen was an active proponent of women's rights. Though Gleditsch graduated from high school in 1895 with the highest marks, the matriculation examination, the prerequisite for university entrance, was barred to women students. As a result, Gleditsch began work as a pharmacy assistant. The work involved a significant amount of chemistry, and she worked toward a nonacademic degree in chemistry and pharmacology, which she obtained in 1902. With the encouragement of Dr. Eyvind Bodtker, who became her friend and mentor, she moved to Oslo and struggled, penniless, to continue her education. From her studies, she wrote a paper on organic chemistry that, on Bodtker's advice, was translated into French and submitted to a French journal, which accepted it for publication.[57]

Gleditsch confided to Bodtker her dream of working with Marie Curie. While Gleditsch looked after the laboratory, Bodtker traveled to Paris to ask Curie to accept her. Initially, Curie turned down his request as she said that there was no room for any more assistants. Bodtker persisted, arguing that Gleditsch was so small that she would take up very little room (her height was only 154 cm). To reinforce his arguments, Bodtker showed Gleditsch's publication to Curie. Curie relented, probably because most of her researchers were physicists and she had a real need for a chemist. Gleditsch obtained a grant from the Norwegian/Swedish Dowager Queen Josephine's legacy, and Curie offered to exempt Gleditsch from the fee that was paid by all who wished to work in her laboratory: "If you would take on this work [recrystallization of barium and radium salts], which will only take up part of your time, and which is of general benefit for the laboratory, I could exempt you from the fees because of these favors. You can at the same time work on another problem of greater interest, which could lead to new results."[58] Gleditsch arrived in Paris in October 1907, spending much of her time performing the tedious separation of radium from the other elements. The special project that was assigned to her was the repetition of experiments by British scientists that claimed exposure of copper salts to radium "emanation" [radon gas] transformed the copper into lithium. Curie and Gleditsch showed that no such transformation occurred.[59] Following completion of this work, Gleditsch embarked upon a study of the relationship between uranium and radium content in minerals.[60]

Gleditsch left Paris in 1911, having been granted a degree from the University of Paris on the basis of her research. Returning to Norway, she was given a Fellowship at the University of Oslo, which entailed lecturing on radioactivity and supervising laboratory work. In 1913, Gleditsch was awarded a scholarship from the American–Scandinavian Foundation to work in the United States. She wrote to Theodore Lyman at Harvard and Bertram Boltwood at Yale for permission to work in their laboratories. Lyman replied that no woman had ever set foot in the physics laboratory at Harvard and implied that he intended to continue the tradition. Boltwood wrote Ernest Rutherford about her letter: "I have written to her and tried to ward her off, but as the letter was unnecessarily delayed in being forwarded to me, I am afraid she will be in New York before I get there. Tell Mrs. Rutherford that a silver fruit dish will make a very nice wedding present!!!"[61] In other words, like many of his contemporaries, he saw women scientists as available marriage partners rather than as enthusiastic researchers.

Presumably Gleditsch decided to journey across the Atlantic anyway, for she did venture to New Haven where a reluctant Boltwood found her research space. It was at Yale that Gleditsch made her most important research contribution, the measurement of the precise half-life of radium. This was the key number in the study of radioactivity, as the science historian Lawrence Badash has commented:

> Probably the most significant constant in radioactivity studies was the half-life of radium, the time required for half its amount to transform into emanation [radon]. As the only highly active radioelement that could be prepared in pure and reasonably large quantities, with a half-life long enough to consider those quantities constant for most practical purposes, radium was regarded as the constant in the field.[62]

Her work gained her the respect of American scientists. Lyman offered her a place as a guest in his laboratory, and the President of the American Chemical Society, Theodore W. Richards, invited her to visit him. In recognition of her contributions to the study of radioactivity, Smith College in Massachusetts awarded her an honorary doctorate in 1914. When she left Yale to return to Norway, even Boltwood had mellowed in his views, writing to her to express his enjoyment of having her work in the laboratory and of sharing their common scientific interests.[63] He also helped her publish her results in the prestigious *American Journal of Science*.[64]

Arguably her most crucial contribution to chemistry came in a more indirect way as a result of her meeting with Richards, who was recognized as the leading analytical chemist of the time.[65] Traditional chemists had assumed that atomic weight was a fundamental property of an element, but the British chemist Frederick Soddy had proposed that atoms of the same element could differ in their atomic weight.[66] Conclusive evidence of this was hard to find, for analytical errors were usually greater than measured differences. Otto Hönigschmid and Stefanie Horovitz (*see* discussion later in this chapter) had skillfully produced results that strongly indicated lead from radioactive sources had a significantly lower atomic weight than that from "normal" lead. Among the chemical fraternity, Richards had such a reputation for accuracy and precision that his measurements would settle the question beyond doubt. Gleditsch collected a variety of Norwegian minerals, extracted the lead salts from them, and sent the salts to

Richards. Though Richards received lead samples from several sources, Gleditsch's samples were those that contained lead with atomic mass significantly different from that of "ordinary" lead.[67] From this result, the existence of isotopes could no longer be questioned.[68]

In 1916, the University of Oslo finally recognized her talent by appointing her to a non-tenure-track faculty position, an advance over the poorly paid research fellowship that she had held until then. The same year, she received a plea from Marie Curie to return to Paris to help revive radium production.[69] This she did, though part of the journey involved a ferry across the dangerous waters of the North Sea where German warships prowled. With her task completed, she returned to Norway and spent the summer of 1917 in Stockholm, Sweden, completing a book on radioactivity[70] that she coauthored with Eva Ramstedt. Ramstedt, a Swedish chemist, had befriended Gleditsch when they were both working with Curie in 1910–1911[71], and they remained close friends throughout their lives. The year 1917 also marked the election of Gleditsch to the Academy of Science, Oslo; she was only the second woman to receive this recognition.

Feeling geographically isolated in Norway, Gleditsch visited and worked in Paris several times in the 1920s, often supervising radium production, and once supervising the Curie research group while Marie Curie visited South America. She was well-known among the other researchers in radioactivity and detoured through England to visit Ernest Rutherford and Frederick Soddy on her journeys between Oslo and Paris. Her research continued on the study of isotopes, some of which was performed with her sister, Liv Gleditsch, who was also a chemist. Ellen Gleditsch wrote a second book, this one on isotopes. The demand for this book was such that it had to be reprinted within a year and an English translation was produced.[72]

In 1929, Gleditsch was appointed professor of chemistry. This was a hard-fought battle by her supporters against the many academics who were opposed to the principle of a woman professor. She built up her own research group and also played a leadership role with the International Federation of University Women. When the Second World War started in 1939 and Norway was occupied by German forces, Gleditsch, at the age of 61, became active with the Resistance. She, like her sister Liv and brother Adler, was arrested, though in her case, only briefly. After the war, she continued research and received many honors for her work, including an honorary doctorate at the age of 83 from the Sorbonne, Paris; she was the first woman to receive this award. She lived to the age of 89, an amazing feat considering the radiation to which she must have been exposed in her early years.

Catherine Chamié (1888–1950)

Gleditsch was one of many researchers who spent one or two years at the Institut de Radium. In fact, as we have mentioned, at any one time there were as many as 30 or 40 researchers working there. Yet there were two women who worked at the Institut de Radium for many years: Irène Joliot-Curie, Marie's daughter; and Catherine Chamié. Chamié did not herself make any "great" discoveries, yet Curie depended upon Chamié for the smooth running of the laboratory.

Chamié was born in Odessa, Russia,[73] the daughter of Antoine Chamié, a Franco–Syrian notary, and Hélène Golovkine. At the time she finished grade school, women were not admitted to Russian universities. Thus at the age of 19, she journeyed across Europe to attend the University of Geneva, Switzerland, and by 1913 she had obtained a doctor of science degree in electrical physics. Upon her return to Russia, she became a research worker in the physics laboratory of the University of Petrograd. When the First World War started, she volunteered as a nurse. Following the withdrawal of Russia from the War in 1917, Chamié worked at the University of Odessa on solutions to differential equations. In 1919, race riots in Odessa forced the entire Franco–Russian community to flee the country, and her family arrived penniless at a refugee camp in Switzerland. After five months there, Catherine Chamié set out for Paris where she obtained employment as a tutor and took courses on radioactivity. She wrote a pleading letter to Marie Curie to ask to be allowed to work in the Curie laboratory. Curie, impressed by the glowing references from Chamié's former professors, agreed to find space for her.

Over the following decades, Chamié published papers on a wide range of topics in radioactivity, the most significant being the discovery of what was called later by other scientists "the Chamié effect." When the chemistry of radioactive elements was studied, the quantities were so small that it was hard to tell if compounds were indeed dissolved in solution or were simply colloidal particles. Chamié exposed photographic film to the radioactive solution, and if the film was uniformly fogged, it had to have been a true solution; conversely, a spotted film indicated colloidal particles.

Though Chamié did pursue independent projects, much of her research was performed collaboratively with other members of the research group, including Marie herself. For example, Chamié recalled the preparation of the isotope actinium X: "The day of work isn't enough for the separation of this element. Madame Curie stays for the evening, without dinner, but the separation...is so slow; so we stay the night, so that the intense source we're preparing won't decay too much. It is already two o'clock in the morning and the last operation is still to be done: centrifuging for an hour."[74]

Of more widespread importance, over the years Chamié became the stalwart of the Institut, classifying radioactive minerals and running the Measurement Service. Her friend and co-researcher for several years, Elizabeth Róna, commented:

> Mlle. C. Chamié, who became my friend, was the custodian of the radium preparations. It was her duty to get the preparations out of a safety box in the morning and return them in the evening. A small cart with some (but not enough) lead bricks around the radium preparation was used, which she pushed to and from the safe. We left together at the end of the work day because we did not live far from each other. But each evening I had to wait outside, because she felt the need to go back to see whether she had really returned the radium preparation.[75]

During Curie's absences, Chamié usually supervised the laboratory. To Chamié, her contribution to the advancement of science was the most important thing in her life; family and financial security were subordinate to the quest for knowledge. Several times Curie had to write letters for financial aid for Chamié, whose existence was always on the poverty line. Chamié continued working at the Institut until 1949, one year before her death, which Róna attributed to excessive exposure to radiation.

Irène Joliot-Curie (1897–1956)

Though Joliot-Curie, like her famous mother, received a Nobel Prize in Chemistry, few recall either her name or her contributions to the study of radioactivity. Joliot-Curie was born in Paris, during the most scientifically productive years of her parents.[76] As a result, Pierre Curie's widowed father, Eugène Curie, usually looked after the young Irène. To assist Eugène, a series of Polish governesses were hired, part of their duties being to ensure that Joliot-Curie became fluent in her mother's language and culture. In addition to mastering academic subjects, Joliot-Curie and her younger sister, Ève, were required to become toughened physically with cycling, swimming, acrobatics, and long walks in harsh weather. When Marie Curie was home, she treated Joliot-Curie the way that she herself had been treated by her mother, never embracing her daughter for fear of infecting her offspring with dangerous bacteria. Joliot-Curie learned early on that her main competition for parental attention was the lab, and she resented every moment that her parents spent there. As an illustration of the unusual family dynamics, when her father was killed, she was not told until after the funeral.

Life was not easy for the young Irène, with the death of her father when she was only nine years old and the second-hand stress of her mother's problems. Her teenage years corresponded to Marie Curie's failure to be elected to the Academy of Sciences, and to the Langevin affair. And while Curie had a lengthy convalescence at a secret location to escape the press, Joliot-Curie was looked after by friends and relatives. Despite the very public controversies, there were some happy times in the Curie family, but Joliot-Curie found much of her pleasure in her studies, particularly mathematics. For example, she wrote to Curie: "The derivatives are coming along alright; the inverse functions are adorable. On the other hand, I can feel my hair stand on end when I think of the theorem of Rolle, and Thomas's formula....."[77]

When the First World War started, the two Curie children were sent with their governess to L'Arcouest, a small fishing village in Brittany, where the Curies had a country home. This was not a pleasant environment, for the local residents accused Joliot-Curie, perhaps because of a Polish accent, of being a German spy. On her seventeenth birthday, Marie agreed that Irène could join her at the Front, working in the X-ray units. The horrors of war that Joliot-Curie was to see over the next four years affected her political attitudes for life. During the shared wartime experience, the mother and daughter developed a close relationship for the first time.

With the War over, Joliot-Curie was hired at the Institut de Radium as assistant to her mother, and while there she taught radiology and continued her studies in mathematics and physics. Joliot-Curie's initial research was on the atomic weight of chlorine,[78] looking for variations similar to those found in lead. However, her doctoral thesis work revolved around a study of the alpha radiation from polonium, for which Curie provided her with the Institut's stock of the element—one of Curie's several interventions on her daughter's behalf that led to accusations of favoritism.[79] During this time, Curie hired a young inexperienced researcher, Frédérick (Fred) Joliot, who was placed under Joliot-Curie's tutelage. A romance soon developed followed by marriage in 1926, Joliot later commenting: "With her cold exterior, forgetting sometimes to say good morning, she did not arouse a feeling of sympathy in the lab. But I discovered in this young woman, whom others saw as a somewhat unpolished block, an extraordinarily sensitive and poetic person who in many ways was the embodiment of what her father had been."[80] They were to have two children, Joliot-Curie working at the research bench until the last possible moment before each birth.

The work of the Joliot-Curies (both Irène and Fred adopting the hyphenated family name upon marriage) was predominantly in the area of particle physics. In several cases, they almost made major discoveries. For example, they studied a mysterious radiation released when boron and beryllium were bombarded with alpha particles, not realizing the radiation was, in fact, the long-sought neutron. Instead, James Chadwick, inspired by the report of the Joliot-Curies, was to be the first to announce the discovery.[81] Then their luck was to change. At the beginning of 1934, they were studying the bombardment of aluminum with alpha particles. One evening, they noticed that the positrons emitted as a result of the bombardment continued to be produced after the alpha source was removed. The only possible explanation was that a new radioactive element had been formed.[82] Fred Joliot-Curie was the instrumentation specialist while Irène Joliot-Curie was the radiochemist. Her task was to isolate this artificially produced element and confirm its existence—a major challenge as the isotope they had formed, phosphorus-30, had a half-life of just over three minutes.

This first recorded artificial synthesis of an element provided the Joliot-Curies with the long-elusive fame, including the Nobel Prize for Chemistry in 1934. Promotion also came rapidly, with Irène Joliot-Curie leading a research group at the University of Paris while Fred Joliot-Curie accepted the chair in nuclear physics at the Collège de France. Continuing their work on the production of elements by particle bombardment, Irène Joliot-Curie, in collaboration with the Yugoslav physicist, Paul Savic, decided in 1938 to investigate one of the products from the bombardment of uranium. This study followed from Enrico Fermi's report that such bombardment led to the formation of previously unknown elements of higher atomic numbers. Joliot-Curie and Savic claimed to have identified a transuranium element which resembled actinium, an element of lower atomic number than uranium.[83] Otto Hahn and Lise Meitner in Germany vigorously disputed this possibility. Yet it was to be Meitner, together with her nephew Otto Frisch, who subsequently explained the Joliot-Curies' observations in terms of nuclear fission, and it was to be the German

Irène (1897–1956) and Frédéric Joliot-Curie in their laboratory, ca. 1935. (French Embassy Press and Information Division, courtesy AIP Emilio Segrè Visual Archives.)

group that gained the fame. Once again, the French researchers had the experimental evidence but failed to make the correct interpretation.

Unlike her mother, Joliot-Curie became active in political causes. Having strong left-wing views, she was made Minister of Science in the short-lived socialist government of 1936. Fred shared his spouse's political activism; thus in 1939, at the start of the Second World War, he joined the Resistance. During the War, Irène Joliot-Curie continued her research work while raising the children. For most of the time, she had little knowledge of Fred's activities or whereabouts. Her health had been deteriorating for many years as a result of her exposure to high levels of radiation, and sojourns in sanitoria became more frequent. By 1944, the Nazis had identified Fred as a major figure in the Resistance, and it became likely that the whole Joliot-Curie family were going to be arrested by the Nazi occupiers. With two guides, Irène Joliot-Curie and the two children trekked across the Jura mountains to the safety of Switzerland. Though very weak from illness, Irène insisted on carrying a heavy textbook of physics across the border.

After the war, the Joliot-Curies led the work on development of the first French nuclear reactor, with Irène Joliot-Curie playing a major role in training uranium prospectors. The membership of Fred Joliot-Curie in the French Communist party and Irène Joliot-Curie's very left-wing, but independent, philosophy was causing them more and more trouble internationally. She was detained at Ellis Island on her third visit to the United States, and in 1954, her application for membership in the American Chemical Society was denied.[84] Following the retirement of André Debierne, she succeeded him as director of the Curie Institut, and she was promoted to full professor at the Sorbonne. However, like her mother, her nomination for membership in the French Academy of Sciences was rejected. Though radiation-related illnesses were continuing to debilitate her, she spoke at international conferences on women's rights and on world peace. She also found time to go hiking in Norway with her mother's friend, Ellen Gleditsch. She died of leukemia at the age of 59.

Marguerite Perey (1909–1975)

The work of most of the women researchers at the Institut de Radium belongs more in the realm of physics than chemistry, but one of the exceptions was Marguerite Perey, who discovered the sixth member of the alkali metals in 1939.

Perey was born in Villenoble, a suburb of Paris, the youngest of five children of an industrialist.[85] She was educated at a school for technicians, from which she graduated with a chemistry diploma in 1929. Hired immediately upon graduation to work at the Institut de Radium, she began work on her first task, which was to chemically separate pure samples of actinium salts. This was a tedious and painstaking task as hundreds of fractional crystallizations were necessary to isolate the very small concentrations of this element from the other components in the radioactive ores. Following Marie Curie's death in 1934, Perey worked on the preparation of pure samples of actinium, a task complicated by the constant radioactive decay of the element itself. During her attempts at purifying actinium, Perey noted a new decay product. This nucleus, which had a half-life of only 22 minutes, resembled cesium in its chemistry. At the time, the sixth member of the alkali metal group, element 87, was still unknown, and the research group quickly realized that this 29-year-old technician had isolated the missing element.[86]

As a result of her success, Debierne and Irène Joliot-Curie encouraged Perey to obtain a formal qualification in science, and in 1946 she received a doctorate in physical sciences from the Sorbonne. Perey spent many years isolating samples of the new element—a complex process as only 1.2% of actinium decayed to give this element. As discoverer, Perey named the element francium, in recognition of her own country, and she studied its chemistry in relation to the other alkali metals.[87] In 1949, she was appointed to the chair of nuclear chemistry at the University of Strasbourg and in 1962, she was the first woman elected to the Academy of Sciences, Paris, the honor denied to her research supervisors, Marie Curie and Irène Joliot-Curie. Like many members of the Curie group, she developed cancer from her work with such highly radioactive sources, dying at the age of 65.

Harriet Brooks (1876–1933)

The French school of radioactivity was focused not only on an individual—Marie Curie—but also at one geographical location—Paris. The British school is associated with the name of Ernest Rutherford and during his career, he moved from Montreal, Canada, to Manchester, England, and then to Cambridge. At Montreal, he had the Canadian, Harriet Brooks, and then the American, Fanny Cook Gates[88] working with him. After Rutherford's move back to England, it was European women such as the Russian, Jadwiga Szmidt,[89] who worked with him. The members of the British school tended to be physicists, which is not surprising as Rutherford considered himself a physicist (though he received a Nobel Prize in Chemistry). Of all his protégées, it was the contributions of Harriet Brooks that were to have the most impact on chemistry. So it is Brooks whom we will choose as a representative of the British School.

Born in Exeter, Ontario, Canada, Brooks was the third of nine children of George Brooks and Elizabeth Worden; her father was a commercial traveler for a flour company.[90] The family moved periodically as a result of her father's peripatetic profession, and by the time that she wished to attend university, they were living in Montreal, Quebec, Canada. As a result, she attended McGill University in that city. She was an outstanding undergraduate student and by a fortunate coincidence, her graduation in 1898 coincided with the arrival of Ernest Rutherford. Rutherford had been lured to Canada by the state-of-the-art physics facilities at McGill, and he invited Brooks to become his graduate student.

With a project on electrical oscillations and an associated M.A. thesis completed, Brooks was assigned a more challenging task, a study of "emanation." Rutherford had given the name "emanation" to a puzzling material released by thorium, and it was unclear whether emanation was a gas, a vapor, or a finely divided solid. Brooks' work showed that emanation was a gas (which we now know to be radon) and that its atomic weight was significantly less than thorium.[91] At the time, the radioactive elements were believed to retain their identity while the radiation was released. Brooks' result was among the first evidence to indicate that a transmutation of one element to another had occurred—an anathema in those days, for chemists were philosophically committed to the immutability of elements.

In 1901, Brooks accepted a Fellowship at Bryn Mawr College, a small women's college in Pennsylvania, to pursue a Ph.D. in physics. While there, she was awarded the prestigious President's Fellowship for graduate study in Europe. This fellowship had been instituted by the President of Bryn Mawr, M. Carey Thomas, who was determined to foster the talent of outstanding women students by sending them to study with the greatest European professors of their chosen specialty. Rutherford arranged for Brooks to work with his former mentor, J. J. Thomson, at the Cavendish Laboratory, Cambridge University. While there, she wrote to Rutherford: "I am afraid I am a terrible bungler in research work, this is so extremely interesting and I am getting along so slowly and so blunderingly with it. I think I shall have to give it up after this year, there are so many other people who can do so much better and in so much less time than I that I do not think my small efforts will ever be missed."[92] Yet this was not the view of Rutherford and others. For example, the McGill University physicist, A. S. Eve, wrote in 1906: "Miss Brooks has published several papers on various radioactive phenomena, and this lady was one of the most successful and industrious workers in the early days of the investigation of the subject."[93] Her lack of self-confidence was, and is, not uncommon among women scientists, as physicist Susan Watson has remarked: "There can be very talented women at the tops of their classes who still feel that their male colleagues are much smarter and that any moment someone's going to reveal how stupid and incompetent they really are."[94]

Upon completing a research year in England, Brooks returned to McGill to work again with Rutherford rather than continue her studies at Bryn Mawr. During this year (1903–04), she made a significant observation: that a nonradioactive plate, placed inside a radioactive container, itself became radioactive. This she interpreted in terms of the volatility of the radioactive substance.[95] Later, Rutherford realized that the correct explanation involved the recoil of the radioactive atom: that is, when a particle is released from an atom, the atom itself will be propelled in the opposite direction, sometimes with enough energy to leave its matrix and attach itself to a new surface. This phenomenon was later utilized to separate daughter products in radioactive decay sequences, and hence identify new isotopes. In the same year, Brooks published some of her results from the Cavendish. This study indicated a rise and fall of radioactivity that could only be interpreted in terms of two successive radioactive changes. This observation laid the groundwork for Rutherford's classic lecture on the concept of radioactive decay sequences, a presentation in which he gave frequent reference to the work of "Miss Brooks".[96]

The following year, Brooks accepted a position at Barnard College, New York, the women's college associated with Columbia University. For the next two years, her time was occupied by teaching advanced physics. In the summer of 1906, she became engaged to a physicist, Bergen Davis of Columbia University. When she told Barnard's Dean, Laura Gill, Gill requested her resignation. In response, Brooks wrote a stirring letter: "I think also it is a duty I owe to my profession and to my sex to show that a woman has a right to the practice of her profession and cannot be condemned to abandon it merely because she marries."[97] The head of physics, Margaret Maltby, pleaded to Gill on Brooks' behalf: "Neither you nor I would like to give up our active professional lives suddenly for domestic life.... I know of no woman to take her place—no one available who has the preparation and the personality and ability to teach, and the skill in physical manipulation that she has."[98] Gill was adamant: "The College cannot afford to have women on the staff to whom the college work is secondary; the College is not willing to stamp with approval a woman to whom self-elected home duties can be secondary."[99]

Though Brooks broke off the engagement, the mental stress caused her to resign from Barnard. That summer, she met Maxim Gorky, the Russian writer and revolutionary, joining his entourage in the Adirondacks. She traveled with the group to Italy, then on her own to Paris where she worked with Marie Curie, extending the research that she had performed with Rutherford. Curie invited her to stay for the 1907–1908 year, but instead she applied for a position at the University of Manchester, England, where Rutherford was about to take over the chair of physics. However, she suddenly withdrew from the competition and married Frank Pitcher. Pitcher had demonstrated physics to her when she was a student at McGill and, for the months prior to marriage, had ardently pursued her.

The Pitchers settled down in Montreal where they raised three children, two of whom died as teenagers. Brooks never did return to science, possibly as a result of her self-image and the expectations of Montreal society, but also her mentor was now across the Atlantic, and radioactivity research at McGill fell into a rapid decline. She died in 1933, probably due to her years of exposure to radioactive materials, particularly radon gas. Eve wrote Rutherford, telling him of Brooks's death. In his reply, Rutherford remarked: "The last time she came to see us about two years ago, one could not but recognize the obvious loss of vitality but this was quite understandable after her family calamities. I have the happiest memories of our friendship in the old days at McGill and the renewal of these during our occasional visits to Montreal. She was a woman of great personal charm as well as of marked intellectual interests."[100]

Stefanie Horovitz (1887–1940)

Horovitz belonged to the Vienna school. This was a much more loose-knit group than those of either Curie or Rutherford. Though Stefan Meyer was nominally the head, individual research leaders carved out their own fields of interest. Nevertheless, the essential point was that here, too, a remarkable number of women were involved in research activities. Many, such as Marietta Blau[101], were pure physicists, while Horovitz was one of the few chemists.

Horovitz was born in Warsaw, the daughter of the artist, Leopold Horovitz.[102] The family moved to Vienna about 1890, where Horovitz entered the University of Vienna in 1907. In 1914, she received a doctorate in organic chemistry for her work on the rearrangement of quinone using sulfuric acid. She joined Otto Hönigschmid at the Radium Institute of Vienna in either 1913 or 1914. Hönigschmid was actually affiliated with the Technical University of Prague, and Horovitz seems to have followed him between the two institutions. At the time, Hönigschmid was one of the many chemists around the world trying to determine precise atomic weight values for elements, and he assigned Horovitz the task of determining the atomic weight of lead from radioactive sources. The only European mine producing large quantities of pitchblende, the source of the lead, was nearby at St. Joachimstal; thus the raw material was at hand. Even then, the separation of pure lead was a long and arduous task, and, when the ultrapure lead had been extracted, the measurements had to be performed to the nearest one hundred thousandth of a gram—an exceptional precision even today. To illustrate their devotion to research, Hönigschmid remarked in a letter to Lise Meitner: "...Miss Horovitz and I worked like coolies. On this beautiful Sunday we are still sitting in the laboratory at 6 o'clock."[103]

There had been claims before that the atomic weight of lead from radioactive sources differed from that of normal lead. However, the data had been rejected as unreliable. Hönigschmid had studied under the famous analytical chemist T. W. Richards at Harvard, and hence the values found by his own student were widely accepted. Horovitz's results were startling to the chemists of the time, for she found the atomic weight of lead from pitchblende to be 206.736 compared to 207.190 for "normal" lead. This difference was enough to demolish the belief in the invariance of atomic weights and provide the first authoritative evidence for the concept of isotopes. The paper on their results, having had such a profound effect on the basis of chemistry, has been cited as a classic of the first half of the twentieth century.[104] As mentioned earlier, Richards himself subsequently repeated this work, definitively establishing the existence of isotopes. Horovitz extended her studies to other sources of lead from radioactive minerals, including one from Norway, almost certainly supplied by Ellen Gleditsch. Her next project was a study of ionium. At the time, ionium was believed to be another radioactive element, but Horovitz showed it to be an isotope of thorium. Thus in one set of results she had disproved the existence of a claimed element and, instead, provided evidence of isotopy for a second.

These were her last publications, and then she disappeared from the scientific records. Unfortunately, Hönigschmid's files were destroyed in the Second World War. The chemist Kasimir Fajans wrote to Elizabeth Róna asking if she knew what became of Horovitz. Then, in a later letter, Fajans wrote:

> You probably have not received any information from Vienna about the fate of Dr. Stephanie Horovitz. I learned about it from a mutual relative at Warzawa. Stephanie moved there after World War I and after her parents had died in Vienna to join her married sister. She was not active in chemistry and both were liquidated by the Nazis in 1940.[105]

Lise Meitner (1878–1968)

Lise Meitner was another member of the Vienna school. Though she considered herself a physicist, several of her discoveries had an impact upon chemistry, and it is those that we will focus on here.

Meitner, born in Vienna, was the third of eight children of Philipp Meitner, a lawyer, and Hedwig Skovran, a talented pianist.[106] During Meitner's youth, education for girls in Vienna terminated at the age of 14, as contemporary educational theory held that girls had learned enough by that age to run a household, raise children, and converse with a husband. Meitner, however, had already decided that she had a passion for physics—what would have seemed an impossible dream. Her father insisted that she first study for a certificate to teach French in girls' finishing schools as a means of ensuring a livelihood. When this was completed, he was prepared to pay for a private tutor to educate her for the university entrance examinations. In two years, Meitner completed the missing eight years of school work required for admission, and she was one of four women permitted to enroll at the University of Vienna in 1901.[107] Throughout her life, she was bitter at the "lost years" caused by the delay in entering university.

Once in university, Meitner announced that she would be so busy with her studies that she would have no time to read newspapers; therefore, she expected her siblings to supply her with a daily bulletin. Being unable to resist the opportunity, her brothers and sisters invented a war and every day they would inform her of the progress of the imaginary battles. That they were able to keep up the pretense for a long time indicates how Meitner's university conversations must have focused solely on academic studies. In 1905, she became the second woman to be awarded a doctoral degree in physics at the University of Vienna.[108] Having been intrigued by the new subject of radioactivity, she then undertook her first research in the field, a study of the deflection of alpha particles.

Meitner felt the need to travel to somewhere more in the forefront of radioactivity research. Marie Curie rejected her application to work in Paris, but Max Planck, though not particularly favorable to women academics, agreed to allow her to attend his lectures at the University of Berlin. At about the same time, Otto Hahn arrived in Berlin. Hahn, who had worked with Rutherford at McGill University, Montreal, was looking for a collaborator and Meitner proved a perfect match in terms of her superiority in theoretical physics to complement his more chemical background. Unfortunately, Hahn worked in the Chemistry Institute, run by the famous organic chemist, Emil Fischer. Fischer forbade women from entering the Institute.

Lise Meitner (1878-1968) on her 21st birthday, 1899. (Courtesy of Churchill College Archives.)

However, he reluctantly allowed Meitner to work in a small basement room originally planned as a carpenter's shop. For five years, Meitner worked under such conditions. From time to time, she would sneak upstairs into forbidden territory and hide under the lecture benches so that she could hear the presentations. During this time, she received no salary, relying on her family for money that she needed for rent and food, the latter being mostly black bread and coffee.

In 1912 the new Kaiser Wilhelm Institute was completed, and Meitner and Hahn moved there to comparatively luxurious surroundings. Planck obtained an assistantship for her to provide a minimal income but her financial situation was little better for, unfortunately, that same year her father died and her family allowance ended. During the First World War, Meitner spent two years as an X-ray nurse in an Austro-Hungarian army hospital. Her leaves were spent back at the research bench, continuing the project that she and Hahn had commenced before the War—the search for the parent element of actinium, that is, the element that decayed to give actinium. With chemical advice by mail from Hahn, then a soldier in the Austro-Hungarian army, she managed to identify this new element, which they named protactinium.[109] At the time, their publication practice seemed to have been that a chemistry article named the chemist, Hahn, first, while a physics article named the physicist, Meitner, first. Thus even though Meitner did most of the work, Hahn was recognized as the lead author of the initial publication.[110] Still, the first paper to mention the name of the new element was in Meitner's name alone.[111]

The postwar republican government in Germany greatly improved the status of women, and in 1926 Meitner became the first woman physics professor in the country. The partnership with Hahn had ended and Meitner ran her own research team. In 1934, however, Meitner asked Hahn to join her to follow up on the reports from Enrico Fermi's group in Italy. Fermi had been bombarding heavy elements with neutrons. He hoped to turn them into new elements beyond uranium in the periodic table, but his results were very confusing. To duplicate and elucidate Fermi's experiments, Meitner needed a chemist and Hahn was the obvious choice. In turn, Hahn brought in Fritz Strassmann, a skilled analytical chemist. Two women completed the team: Clara Lieber from St. Louis, Missouri, and the technician, Irmgard Bohne. By the end of 1937, Meitner and Hahn believed they had identified at least nine new elements while Irène Joliot-Curie in Paris had claimed to have found others. Though Hahn was delighted by all these new elements, Meitner was more puzzled than confident. Unfortunately, politics now intervened. When Hitler came to power in 1933, non-Aryans (i.e., those other than north European Caucasians) were dismissed from their academic posts. As a Jew by birth (though she had converted to Christianity), Meitner was only saved in the early years by the fact that she was Austrian and not German. When the German army occupied Austria and the country was incorporated into the German state, her protection ceased. A Dutch X-ray spectroscopist, Dirk Coster, helped her escape from Germany, a difficult and dangerous task.[112]

Meitner stayed briefly with the physicist, Niels Bohr, in Copenhagen, but accepted a position with Manne Siegbahn at the Nobel Institute for Experimental Physics in Stockholm. This was a mistake, as Meitner's biographer Ruth Sime has commented: "Meitner chose Sweden, however, not because it was familiar—she had closer friends in Holland and England—but because nuclear physics was undeveloped in Sweden and she believed she could be of use. She did not foresee how impossible it would be to duplicate the spirit of her own institute in Berlin, and she did not know that Siegbahn was in fact reluctant to have her."[113] Fortunately, she did not accept a subsequent offer from Holland, as that country was to be overrun by the Axis forces. However, it was very unfortunate that she did not accept a later offer from Cambridge University in England, where her welcome would have been much more hospitable.

Sweden had the one advantage of an overnight mail service between Berlin and Stockholm, and Meitner and Hahn used it to correspond every other day. On 19 December 1938, Hahn wrote a crucial letter to Meitner: "...There is something about the radium isotopes that's so curious that we wanted to first tell it only to you.... Our 'radium isotope' behaves like barium.... Perhaps you can propose some fantastic explanation."[114] Concerned that the Joliot-Curies might be coming to the same discovery, Hahn and Strassmann hurriedly drafted a paper on their findings and had it accepted for publication on 6 January 1939.[115] At the end of 1938, Meitner was visited by her physicist nephew, Otto Frisch, from Denmark. This was to be a momentous meeting which owed its significance, in part, to the backgrounds of the individuals involved. Though both had originated from the University of Vienna, Meitner was a member of the Berlin school of atomic physics while Frisch had been influenced by Niels Bohr, at the Copenhagen school.[116] Bohr had been a proponent of the liquid drop model of the nucleus, a concept first proposed in 1928 by the Russian scientist, George Gamow.

Frisch has described his recollection of the events:

This is where I came in because Lise Meitner was lonely in
Sweden and, as her faithful nephew, I went to visit her at
Christmas. There, in a small hotel in Kungälv near Göte-
borg I found her at breakfast brooding over a letter from
Hahn. I was skeptical about the contents—that barium was
formed from uranium by neutrons—but she kept on with
it. We walked up and down in the snow, I on skis and she
on foot, and gradually the idea took shape that this was no
chipping or cracking of the nucleus but rather a process to
be explained by Bohr's idea that the nucleus was like a liquid drop; such a drop might elongate and divide itself.
Then I worked out the way the electric charge of the
nucleus would diminish the surface tension and found
that it would be down to zero around $Z = 100$ and probably
quite small for uranium. Lise Meitner worked out the
energies that would be available from the mass defect in
such a break-up. She had the mass defect curve pretty
much in her head. It turned out that the electric repulsion
would give them about 200 MeV of energy and that the
mass defect would indeed deliver that energy so that the
process could take place on a purely classical basis without
having to invoke the crossing of a potential barrier, which
of course could never have worked.[117]

Thus all the new claimed post-uranium elements were nothing more
than fission products—isotopes of already known elements. Of equal
importance, they had calculated that the fission process would release
enormous quantities of energy. Unfortunately, with Meitner's lack of
access to equipment, she could not perform the confirmatory experiments. Frisch returned to Copenhagen and successfully performed
these experiments, and the pair sent a note of their proposal for
nuclear fission to the journal *Nature*.[118] Meitner sent a copy of the
manuscript to Hahn, and Hahn and Strassmann quickly identified
one of their products, not as a new transuranium element, but as barium itself. They rushed their findings into print, but made hardly any
mention of Meitner and Frisch.[119] And it was to be Hahn's name that
became linked with the discovery of the fission process.

Hahn was never to acknowledge Meitner's contribution to unrav-
eling the puzzle of nuclear fission, though she initiated the project
and explained the results. Sime has remarked:

> Had Hahn made an effort to set the scientific record
> straight, had he spoken of their long friendship and col-
> laboration, of her leadership of the Berlin team or her

contributions to the fission discovery, it would have
helped. But he did not. In his mind the discovery had
become his, and his alone. In his many interviews he never
spoke of his work with Meitner; not once did he even mention her name.... Meitner and her friends were appalled.[120]

In 1944, Hahn was awarded the Nobel Prize in Chemistry. The decision was not announced at the time because Germans under the
Third Reich were forbidden from accepting Nobel awards. Sime
added: "The decision must have been hasty, however, because in 1945
the chemistry section of the Royal Academy took the highly unusual
step of reconsidering its 1944 award to evaluate the contributions of
others, in particular Meitner and Frisch. After much debate, a slim
majority of the entire academy voted to retain its original decision."[121]

Meitner was nominated several times for a Nobel Prize in Physics,
each time unsuccessfully. The reason that she was unsuccessful seems
to have been the malevolent influence of Siegbahn. Siegbahn, himself a Nobel laureate, was a major influence in Swedish physics and a
member of the selection committee for the Nobel Prize in Physics. In
a letter to Meitner, the Swedish scientist Hans Pettersson definitely
blamed Siegbahn: "[We] are outraged about the one-sidedness of the
distribution of the Nobel Prize. We are certainly glad that Hahn got
the chemistry prize but the physics prize should have gone to you. I
personally am quite sure this would have happened had not Sweden's
most skillful manipulator [Siegbahn] been against it for mysterious
reasons of prestige."[122] As a maximum of three individuals can share
a Nobel Prize, the fairest solution would have been for Hahn and
Strassmann (whose contributions were also ignored) to share the
chemistry prize and for Meitner and Frisch to share the physics prize.
But that was not to be.

Staying in Sweden throughout the Second World War, Meitner
refused to move to the United States to work on the Manhattan
Project to construct a nuclear weapon. To her disgust, Hahn, though
expressing anti-Nazi views, continued to work in Berlin on atomic fission. After the War, they maintained a cordial personal relationship
despite their very different perspectives on the fission discovery and
on the war. Refusing to return to work in Germany, Meitner stayed in
Sweden until 1960 when she retired to Cambridge, England to be
near Frisch. She died just before her ninetieth birthday. Hahn's viewpoint of the fission discovery was widely accepted in Germany and
only recently has Meitner's role been acknowledged.[123] Meitner's
name has finally been immortalized by naming element 109 "meitner-
ium" after her.[124]

Ida Tacke Noddack (1896–1978)

Though Meitner was the first to have conclusive proof of nuclear fission, Ida Noddack, a geochemist, first proposed the concept. Her other claim to fame was the co-discovery of the element rhenium. Yet her German nationalism seems to have been part of the reason for her disappearance from history.[125]

Ida Eva Tacke was born in Lackhausen, near Cologne, Germany, daughter of Adelbert Tacke, a lacquer and varnish manufacturer, and Hedwig Danner.[126] She studied chemistry and metallurgy at the Charlottenburg Technical University, taking an industrial chemistry position on graduation. In early 1925, she joined the Imperial Physico-Technical Research Office, a government laboratory in Berlin, and started working with Walter Noddack. They developed an interest in finding the two missing members of Group 7, elements 43 and 75. Previous researchers had focused on manganese ores, but Tacke and Noddack noted that elements of the second and third transition series were more often found with other members of those rows rather than with the first transition series analog. Thus they looked for traces of these new elements in ores of molybdenum, tungsten, ruthenium, and osmium. In June 1925, they identified element 75, which they named rhenium.[127] In this research, it is clear that Tacke played the leading role.

Though Tacke became famous in Germany, being the first woman ever to address the Society of German Chemists, Tacke and Noddack were to be dogged by criticism elsewhere. First, they also claimed to have discovered element 43, which they called masurium, though no one else could reproduce their results.[128] Second, the names they had chosen for the two elements could be interpreted as references to German geography, or the limits of Allied advance in the First World War: the Rhine to the west and the Masurian marshes to the east (these marshes were also the site of a battle in which the German army annihilated the Russian forces). This attempt at using chemistry to promote nationalism was unforgivable to many chemists.

In 1926, Tacke married Noddack, becoming a "wife-chemist." No longer did she have any formal status, only unpaid access to a research laboratory arranged by her husband and his associates. This lack of formal status contributed to the ignoring of her ideas. In 1934, she read of Fermi's claim to have produced new heavy elements, including element 93. This she did not believe, responding: "When heavy nuclei are bombarded by neutrons, it would seem reasonable to conceive that they break down into numerous large fragments which are isotopes of known elements but are not neighbors of the bombarded elements."[129]

This view was dismissed, as Noddack later recalled: "When in 1935 or 1936 my husband suggested to Hahn by word of mouth that he should make some reference, in his lectures and publications, to my criticism of Fermi's experiments, Hahn answered that he did not want to make me look ridiculous as my assumption of the bursting of the uranium nucleus into larger fragments was really absurd."[130] Ida Noddack's wife-chemist status and the publication of her work only in German journals were factors that prevented her suggestion of uranium fission from receiving any serious consideration outside of Germany.

In 1935, the Noddacks moved to the University of Freiburg, then in 1942 to the University of Strasbourg in occupied France; in both cases only Walter Noddack had a formal academic appointment. After the war, suspicions were raised of the Noddacks's tolerance towards the Nazi regime. Between 1944 and 1956, Ida Noddack was unemployed. During this period, the Noddacks spent some time in Turkey. Ida Noddack resumed geochemical research in 1956, which she continued until her retirement in 1966. She died at the age of 82.

Maria Goeppert-Mayer (1906–1972)

Goeppert-Mayer was the second of the women Nobel laureates of atomic science. Though her life starts beyond our normal limit of the nineteenth century, we would be remiss to exclude her simply on this basis.

Maria Gertrude Göppert was born in Upper Silesia (then a region of Germany, now part of Poland) to Friedrich Göppert and Maria Wolff.[131] She was an only child, and her pediatrician father strongly encouraged her academic interests. This was easy, for in 1910 the family had moved to Göttingen, a major intellectual center. She entered the famous University of Göttingen, planning to major in mathematics and then teach high school mathematics, but she became fascinated by the exciting world of atomic physics. In 1928, she obtained a scholarship to spend a term at Cambridge University where she attended Rutherford's lectures and gained fluency in the English language. Returning to Göttingen, she continued her studies toward a doctorate in quantum mechanics. The following year, the family rented a room to Joseph Mayer, a student from Berkeley who was studying for a year at the University. A romance developed and Mayer and Göppert were married in 1930, the same year that Goeppert-Mayer (having Anglicized her German name) completed her degree.

Following her graduation, the couple moved to Baltimore, Maryland, where Mayer had been appointed associate professor of chemistry at Johns Hopkins University. Though Goeppert-Mayer had an outstanding background in quantum mechanics, the antinepotism regulations forbade offering her a formal position at Johns Hopkins. Her biographer, Joan Dash, has commented that she was regarded as an "...academic wife, whose husband has a fairly secure position, so that the university considers her captive, who can be hired with no guarantees, for as little as possible, and dropped whenever it becomes expedient."[132]

In 1933, her first child, Marianne, was born, and Goeppert-Mayer cut back on her research. The following year, she employed a nurse and returned to research with renewed intensity. Both Marianne and Peter (born in 1938) came to resent their mother's focus on her research work. Goeppert-Mayer and Mayer started work on a textbook of statistical mechanics.[133] Unfortunately, Mayer was fired from Johns Hopkins in 1939. The official reason was that he had not attracted enough graduate students, but other reasons surfaced later, such as a desire by the administration to purge Communists, Jews, Catholics, women, and foreigners.[134] Goeppert-Mayer herself felt that her active presence in the department had been a major factor. Mayer accepted an associate professor position at Columbia University, but Goeppert-Mayer was again excluded from any formal appointment. The chair of the Columbia department of physics made it clear that Goeppert-Mayer was unwelcome in his department, but the chair of chemistry, Harold Urey, came to the rescue and provided her with a lectureship in chemistry. Later, in 1941, Goeppert-Mayer obtained a part-time teaching position at Sarah Lawrence College in Bronxville, New York. The following year, Urey arranged for her to join his secret isotope-separation group developing the atomic bomb, and at last Goeppert-Mayer was able to have a lab of her own. The stress of the war work contributed to her increasing consumption of alcohol and cigarettes, and though she tried to give up her nicotine addiction, she would sometimes smoke continuously, even three or four cigarettes at a time.

After the end of the War, the whole research group moved to the University of Chicago, where she held the rank of "voluntary professor", another unpaid position. This did not concern her, for she was extremely happy at Chicago. One of her former students from the Johns Hopkins days, Bob Sachs, had become head of the theory division of the Argonne National Laboratory, and he offered her a part-time paid position there where she worked with Edward Teller on the

Maria Goeppert-Mayer (1906–1972) with Victor Weisskopf (left) and Max Born. (Courtesy of American Institute of Physics Emilio Segrè Visual Archives.)

origins of the chemical elements. Teller lost interest, but she perse-
vered. As part of the study, she was examining the values of isotopic
abundances in the universe. In general, with increase in atomic number, the abundance of the elements decreases. However, she noticed
that there were certain isotopes that were more abundant than the
trend would predict. These isotopes had either 50 or 82 protons, or
50 or 82 neutrons, numbers that Goeppert-Mayer referred to as
"magic". Subsequent studies led her to believe that such isotopes, as
well as those with 2, 8, 20, 28, and 126 protons and/or neutrons, possessed exceptional stability.[135]

Researchers before her had tried to develop a model of nuclear
structure similar to that of electron structure, but without success.
Goeppert-Mayer realized that an analogous model *could* be developed
if l, the orbital angular momentum number, and s, the spin angular
momentum number, were considered to interact. The orbital and
spin quantum numbers could reinforce or oppose each other. As a
result, this spin–orbit coupling would lead to a splitting of the energy
levels. The same phenomenon exists for electrons but is of much
more importance for the nuclear particles. It was the splitting due to
spin–orbit coupling that resulted in the shell filling being different
for protons and neutrons than that for electrons. The shell model has
since proved extremely useful in understanding isotope abundances
and nuclear stabilities.[136]

Meanwhile, Hans Jensen was making the same discovery at the
University of Heidelberg in Germany. After they had published their
work independently, Goeppert-Mayer and Jensen collaborated on the
classic text describing the shell model: *Elementary Theory of Nuclear
Shell Structure*.[137] They were jointly awarded the Nobel Prize for Physics for this work in 1963. In 1959, the Mayers moved to the University
of California at La Jolla, and only then, at age 53, did Goeppert-Mayer
receive a full-paid professorship. Unfortunately, the arrival in California coincided with her worsening health: first, a mild stroke, and then
heart problems, the latter causing her death in San Diego in 1972.

The Change in the Nature of Atomic Research

After the First World War, in most institutions a decreasing proportion of women entered the field of radioactivity and those that did
were primarily employed in technician roles (such as Perey).[138] For
example, when James Chadwick visited Vienna in 1927, he found that
the routine measurements of evidence of radioactive products were

being performed exclusively by women. The reason given by Stefan Meyer's associates was that "...they [the women] could concentrate on the task more intensely than men, having little on their minds anyway, and by Slavic women because their large, round eyes were best suited for counting."[139]

We will discuss the general reasons for the decline of women's participation in science in Chapter 10, but there were some special factors that were involved in the field of radioactivity. In those early years, the study of radioactivity bridged chemistry and physics, but with the discovery of isotopes and of transmutation, the chemical problems were solved.[140] As a result the field of radioactivity itself transmuted into atomic physics. No longer the "orphan", atomic science attracted those seeking scientific fame and fortune.[141] As the physicist J. C. Slater commented: "Here was a decade [the 1920s] in which the most momentous discoveries in a century were being made, and here were fifty or more ambitious young men, entering a field with a smaller number of older and very distinguished workers, all trying to be in on the exciting discoveries that all were convinced were going to be made."[142]

But there is a second reason for the disappearance of women. The United States became one of the new foci, but this new center had no tradition of encouraging women in radioactivity.[143] For example, Gleditsch had, initially, a very cold reception at Harvard and Yale. This attitude continued to be prevalent, as was apparent by the peripheral roles of women scientists in the 1940s Manhattan Project,[144] and it continued into the 1950s. Fay Ajzenberg-Selove, who was working on the energy levels in the nucleus of low-atomic-number atoms, remarked: "In 1955 the life of a woman physicist was very rough. When she heard that I was getting married, Maria Goeppert-Mayer, who saw me at a meeting, invited me to her room and poured me a stiff slug of whiskey. She told me that while it was hard to be a woman physicist, it was nearly impossible to be a married woman physicist."[145] And the attitudes specifically in atomic physics were little better in the 1980s, as the physicist Sharon Traweek observed: "The traits required for gaining entry into this exclusive community—aggressive individualism, haughty self-confidence, and a sharp competitive edge—are traits typically defined as masculine by our society."[146] She concluded that "extreme masculinity" was the only basis for success—a far cry from the collaborative atmospheres of the early laboratories that once were so favorable to women's participation in the field of atomic science.

being performed exclusively by women. The reason given by Stefan Meyer's associates was that "...they [the women] could concentrate on the task more intensely than men, having little on their minds anyway, and by Slavic women because their large, round eyes were best suited for counting."[139]

We will discuss the general reasons for the decline of women's participation in science in Chapter 10, but there were some special factors that were involved in the field of radioactivity. In those early years, the study of radioactivity bridged chemistry and physics, but with the discovery of isotopes and of transmutation, the chemical problems were solved.[140] As a result the field of radioactivity itself transmuted into atomic physics. No longer the "orphan", atomic science attracted those seeking scientific fame and fortune.[141] As the physicist J. C. Slater commented: "Here was a decade [the 1920s] in which the most momentous discoveries in a century were being made, and here were fifty or more ambitious young men, entering a field with a smaller number of older and very distinguished workers, all trying to be in on the exciting discoveries that all were convinced were going to be made."[142]

But there is a second reason for the disappearance of women. The United States became one of the new foci, but this new center had no tradition of encouraging women in radioactivity.[143] For example, Gleditsch had, initially, a very cold reception at Harvard and Yale. This attitude continued to be prevalent, as was apparent by the peripheral roles of women scientists in the 1940s Manhattan Project,[144] and it continued into the 1950s. Fay Ajzenberg-Selove, who was working on the energy levels in the nucleus of low-atomic-number atoms, remarked: "In 1955 the life of a woman physicist was very rough. When she heard that I was getting married, Maria Goeppert-Mayer, who saw me at a meeting, invited me to her room and poured me a stiff slug of whiskey. She told me that while it was hard to be a woman physicist, it was nearly impossible to be a married woman physicist."[145] And the attitudes specifically in atomic physics were little better in the 1980s, as the physicist Sharon Traweek observed: "The traits required for gaining entry into this exclusive community—aggressive individualism, haughty self-confidence, and a sharp competitive edge—are traits typically defined as masculine by our society."[146] She concluded that "extreme masculinity" was the only basis for success—a far cry from the collaborative atmospheres of the early laboratories that once were so favorable to women's participation in the field of atomic science.

Chapter 7

Women in Biochemistry

The Origins of Biochemistry

The first applications of chemistry to the life sciences date back several centuries. The concepts of photosynthesis and respiration, for example, owe much to the work of Priestley and Lavoisier in the 1770s,[1] and the general study of biological components such as albuminous matter (proteins), sugars (carbohydrates), and fatty substances (lipids) developed around the beginning of the nineteenth century. Yet philosophically, early biochemistry was a type of science very different from modern biochemistry. In particular, substances derived from living matter were supposed to possess a *vital force*, a property that distinguished these compounds from those derived from inanimate sources. This idea was first proposed by Friedrich Casimir Medicus, and it became generally accepted by the end of the eighteenth century. To illustrate how fundamental this principle had become, the first article in the inaugural issue of the prestigious journal *Archiv für Physiologie* was devoted to a thorough elucidation of the principles of this vital force.[2] Vitalism remained the basic principle of the life sciences through most of the early nineteenth century,[3] with a

135

landmark experiment in its overthrow being the chemical synthesis of urea by Friedrich Wöhler in 1828.[4] The reception of this experiment at the time[5] and its significance for the fall of vitalism[6] are still a matter for debate, but the 1830s do seem to provide a turning point in the chemical study of biological materials.[7]

Nevertheless, like the studies of radioactivity and X-ray crystallography, modern biochemistry was a science that started in the early part of the twentieth century. The science historian Robert Kohler has remarked: "Biochemistry is one of those fascinating but problematic 'new sciences' that have appeared with some regularity in the history of science. It came quite suddenly on the scene in the early years of this century, with a new name and intimations of new insights into the nature of life processes."[8] Arguably the first significant contribution to this new field was written in 1901 by the German scientist, Franz Hofmeister. In his paper "The Chemical Organization of the Cell,"[9] Hofmeister reasoned that, within each living cell, a coordinated multitude of chemical processes occur. But it was the address in 1913 by the physiological chemist, Frederick Gowland Hopkins, on the "dynamic side of biochemistry,"[10] that really defined the science as the chemical study of biological processes. Even then, partially because of the First World War, it was between the 1920s and 1940s that biochemistry, particularly the study of enzyme processes, expanded into the modern science that we know today.

The Mentors

Biochemistry was the third field that contained a high proportion of women. As we mentioned in Chapter 5, Rossiter proposed that women were often to be found in rapidly growing fields.[11] In the case of biochemistry, Mary Creese supported Rossiter's hypothesis, and she noted that researchers could enter the field from a wide range of backgrounds: "The entry paths and entry qualifications of its practitioners were not well defined. This lack of prestige, due to the slowness of academic chemists to recognise the full power and potential of research in the field, offers one explanation for its relative openness to women."[12]

Yet the overriding factor seems to have been the encouragement of their women protégées by the leading researchers. As we mentioned earlier, the founder of modern biochemistry was F. Gowland Hopkins, later Nobel laureate, and it was he who initiated the female-friendly nature of the field. Creese has commented:

At the time when there were practically no women research workers in any of the other university departments at Cambridge, Hopkins gave them places in his, despite the criticism which this brought him. Even in the 1920s and 1930s, when, as a Nobel laureate with a worldwide reputation he received hundreds of applications for places in his laboratory, nearly half of the posts in his Department went to women scientists.[13]

The personal style of Hopkins was also an important influence. Dorothy M. Needham, one of Hopkins's students, contended that Hopkins provided valuable moral support and that he regarded his students as fellow researchers rather than as underlings in a research empire.[14] Another of his former students, Malcolm Dixon, explained:

> Hopkins was one of the kindest and most lovable of men.... He never made an unkind or irritable remark, [and] though he could be critical on occasion, it was always with a courtesy that left no sting. He had great charm of manner and was invariably courteous, even to the least important of us. He was always ready to talk over our work and ideas, and somehow contributed to make us feel by the way that he listened to us that our ideas were extremely interesting and our work important.[15]

Dixon also mentions the importance of the excitement of the field, another factor that attracted women into specific areas of research: "To work there was to feel the thrill and sense of adventure in penetrating into the secrets of living matter, and life was never dull."[15]

In the United States, Lafayette Mendel at Yale University provided the mentoring role for women biochemists. From his appointment in 1896 until his death in 1935, Mendel trained about 124 Ph.D.'s in biochemistry, of whom 48 were women.[16] Icie Macy Hoobler, one of his former students, felt Mendel's support and encouragement had been vital for her morale. She added: "He had such a delightful and helpful way of starting you off on the right path and then urging and stimulating to keep you on the move forward."[17]

Hopkins and Mendel maintained a very friendly correspondence throughout their lives,[18] even though their groups were in competition to find the "beneficial factor" in milk—identified by Hopkins as what we now call vitamins.[19] Both leaders engendered in their research students a social cohesiveness that was quite remarkable. Students of Hopkins set up "Hoppie Societies" wherever they were, and held periodic meetings. For example, the Hoppie Society of America met from 1934 until (at least) some time in the 1940s and there was even a Hoppie Club of Leningrad.[20] Of Mendel, Rossiter has commented, "Most Mendel students remained lifelong friends, and continued for years the tradition of a Yale dinner at national meetings."[21] The women of Mendel's group were particularly tightly knit, as Rossiter added: "Thus, to a certain extent Mendel's women formed a kind of "young-girl network" in physiological chemistry, starting in or even before graduate school and continuing for decades, as they corresponded, saw each other at national meetings, hired (or succeeded) each other for temporary or permanent positions, and in later years, nominated one another for professional awards and elective offices."[22]

It was the American women who gained the greatest acclaim, and hence they will be the focus for our selection of individual biographies of women biochemists. To limit the range of our coverage, we have avoided straying into a discussion of the American women physiologists, a topic ably reviewed by Toby Appel.[23]

Biographies

Icie Macy Hoobler (1892–1984)

Of the graduates of Mendel, Hoobler must be regarded as the most famous.[24] Icie Macy was born on 23 July 1892 into the Macy farm family near Gallatin, Missouri. Though her parents had little schooling, they were determined that their four children (Macy had an older brother and sister, and a younger brother) would have the best available education. In 1907, Macy joined her sister at the Central College for Women in Lexington, Missouri, where, following her father's wishes, she registered for a degree in music. Though she practiced piano day and night, after three years she decided that her talents and interests were not in that direction. Macy felt this initial failure created an inferiority complex that plagued her for years afterwards. She transferred into an English program from which she graduated in 1914. It was at the Central College that she met her first mentor, the biology teacher Lily Egbert, who instilled in Macy an interest in science. Macy commented: "[Egbert] was not only gifted intellectually and exceedingly well-informed, but also a creative science teacher who captured and held the interests of her students. Her profound influence instilled a love of learning, an appreciation of the intricacies and beauty in nature, and the desire to pursue science in the service of mankind."[25]

When her parents asked Macy what she planned to do with her life, she replied that she wished to pursue a science career and that she intended to apply to the University of Chicago, a school renowned for its chemistry and physics program. Having had a somewhat sheltered upbringing, her parents argued for her to initially attend Randolph-Macon Woman's College in Virginia. This she did and it was at Randolph-Macon that she met her second mentor, the acting head of chemistry, Mary Lura Sherrill (1888–1968).[26] At that time, Sherrill was beginning her own Ph.D. at the University of Chicago, whose chemistry department had already gained a reputation as congenial to women, largely due to the influence of Julius Stieglitz.[27] Sherrill convinced Macy to continue her studies in chemistry at Chicago; thus after one year, Macy tranferred there to take a major in chemistry and a minor in physics.

Macy's year at Chicago, under the wing of Stieglitz, filled her full of excitement with the potential for new scientific discoveries, and the cosmopolitan atmosphere enabled her to develop socially and culturally. Prior to her graduation in 1916, Stieglitz asked her to join him as a graduate student, but he suggested that she would benefit by a year or two teaching in a small school first. He found her a position teaching inorganic chemistry at the University of Colorado at Boulder. She was the only woman on the chemistry faculty, and it proved a rough and traumatic experience. In particular, many of the male students made life difficult for this young woman. While at Boulder, she registered for an M.S. degree, her first research project being the development of a more sensitive test for cyanide in autopsies. (Cyanide being the most common method of suicide at the time.) This research provided her first publication.[28] Her second project was to improve the method for the extraction of tungstic acid from tungsten ores, her supervisor having holdings in some Colorado tungsten mines. Though Macy completed the work, she refused to submit it to complete her thesis requirements as she felt that she had been used for the professor's personal gain and that the project was unworthy of the thesis.

She stayed a second year at Colorado, during which time she was invited to become an assistant in the physiological chemistry course in the medical school. She was under the direction of Robert C. Lewis, a graduate of Mendel. Lewis was to become her third inspiring teacher and mentor, and it was Lewis who persuaded her that the acceptability and career potential for women in physiological chemistry and nutrition was much greater than in mainstream chemistry. Macy wrote Stieglitz asking for his advice and guidance. Stieglitz replied that he regretted losing her, but urged her to follow Lewis's advice. Lewis and Mendel arranged for her to be accepted into Yale Graduate School, though the administration at the University of Colorado persuaded Macy to take the oral examination and thus complete her M.S. requirements before leaving there.

Macy arrived at Yale in 1918, and she found life with Mendel a wonderful adventure:

> It was a magnificent experience to be guided by the master mind of Professor Mendel through the intricacies of choosing a research project, planning and selecting techniques to be used, and assembling equipment for the task. It was this close fellowship with the scholarly professor that taught me the real love of searching for scientific truth, independence of thought, and respect for the guidance to be found in the history of science. I grasped the practical approach to scientific problems and a deep concern for human welfare. I learned the importance of objective criticism, growth by constructive criticism, and that public service was a pleasant and fruitful obligation.[29]

For her research work, Macy studied cotton seed meal injury among farm animals. This debilitating illness was believed to be due to starvation or a lack of vitamins, but Macy showed that it was a result of the toxic compound, gossypol, that was found in cotton.[30]

Mendel took pride in the intellectual development and career placement of each of his students. Upon her graduation in 1920, Mendel called Macy to his office and informed her of the interviews that he had scheduled. One of the job offers was an assistant biochemist position at a Pittsburgh hospital. Though this was the least prestigious of the offers, she chose it for the following reason:

> Previously, Professor Mendel had given a provocative lecture on the dairy and milk industry in which he pointed out the dearth of information on human beings.... Then he remarked, "This health field is so important to mankind and it is one to which women scientists have so much to contribute." He looked directly at me! I got his message and left the lecture hall inspired and determined that the health of mothers, infants, and children was to be my first priority in research in the future.[31]

Her initial period at the hospital was not pleasant. Only restrooms for men were provided in the laboratory complex, and she had to use a public restroom in a building some distance away. Her attempts to minimize the number of journeys caused her to develop acute nephritis. She was not allowed to eat in the all-male doctor's dining room as a result of her gender, but neither was she allowed to eat in the nurse's dining room as a result of her status. Thus she had to eat lunch with the nonprofessional staff. This was a major problem as the limited opening hours of this facility were hard to match to her own busy schedule. When her concerns were not addressed, she submitted a letter of resignation. This letter reached the President of the Board of Trustees. His anger at Macy's plight was only increased when he discovered that the reason for her absence from the President's annual staff banquet was that the laboratory chief had deliberately failed to inform her of the event. The President was insistent that Macy should have all the rights of the male employees, and conditions subsequently improved for her.

During this year, Mendel asked Macy to present a paper on her doctoral thesis at a conference in Chicago. It was there that she met Agnes Fay Morgan (1884–1968), head of the household science department of the University of California at Berkeley. Morgan, yet another doctoral student of Stieglitz[32], invited Macy to California to recover her health and to teach food chemistry. At Berkeley, Macy took on many tasks including setting up her first research program. Mendel visited Berkeley in 1923, and he advised her to apply for a position at the Merrill-Palmer School for Motherhood and Child Development in Detroit. Macy did so, and she accepted a post as director of the Nutrition Research Project.

It was her work as director of the Project, later run by the Children's Fund of Michigan, that was to provide her fame as a scientist. During the next 31 years, Macy directed research in such areas as women's metabolism during the reproductive cycle[33]; nutrition for women[34]; composition and secretion of human milk; infant growth and development; nutrition and chemical growth in childhood; evaluation and advice on food purchasing and preparation, and nutritional status in children's institutions in Michigan; and blood studies in health and disease. Over that period, about 150 researchers coauthored nearly 300 publications from her research group. Her most popular book, *Chemical Anthropology*, was devoted to a long-term study of metabolism, diet, and growth in children.[35]

In the 1920s, woman chemists still had an uphill battle for acceptance, as one of Macy's experiences illustrated:

> In 1923 I was the recipient of an appointment in Detroit, Michigan, and as was my custom, I immediately started

attending the monthly meetings of the Detroit Section of the American Chemical Society and found only men in attendance. My associate and I sat on one side of the room and the men on the other with no communication between, leaving the impression that the women chemists were interlopers.... After a few months, Mrs. Lindberg...joined us. Now there were three women sitting on one side of the room and across the aisle there were about twenty-five male chemists.[36]

Finally, one male chemist broke the ice and "[t]hereafter we were received and welcomed by all members as scientists not women."[36] In fact, she was elected chair of the local section in 1930.

During the early years in Michigan, Macy raised the two daughters of her older sister, following the sister's death. To cope, Macy began work at 4 A.M. so that she could spend her afternoons at home with the children. In 1938, Macy married B. Raymond Hoobler, a retired Detroit pediatrician. Macy had known Hoobler professionally for a long time and she had obtained many of her samples of human milk from his Mother's Milk Bureau at the Women's Hospital. Unfortunately, happiness outside of her professional work did not last long, with Hoobler dying on their fifth wedding anniversary.

Macy received several honors for her work, and she became the chair of the Division of Biological Chemistry, the first woman to chair a section of the American Chemical Society (ACS). Macy worked at the Merrill-Palmer Institute until retiring in 1959, but she continued as a consultant until 1974. She returned to Missouri in 1982 where she died in her home town of Gallatin in 1984 at age 92.

Mary Engle Pennington (1872–1952)

Pennington, another of Mendel's protégées, was the elder daughter of Henry Pennington and Sarah B. Molony. She was born in Nashville, Tennessee, but her Quaker family moved back to its roots in Philadelphia soon after she was born.[37] Her interest in chemistry started at age 12, when she read a library book on medicinal chemistry, then walked over to the University of Pennsylvania and asked for the unfamiliar words to be explained to her. She was told to return when she was older, and this she did indeed do, being admitted in 1890 upon graduation from high school. Pennington studied biology, chemistry, and hygiene but, though she completed the requirements for a B.S. in 1892, only received a certificate of proficiency, for at that time the University did not grant bachelor's degrees to women. However, she was permitted to enroll in graduate school where she completed a doctorate in 1895 on compounds of columbium (niobium) and tantalum[38] with the famous chemist, Edgar Fahs Smith.

Pennington remained at the University of Pennsylvania for two more years, working as a fellow in chemical botany. Then, in 1897, she spent a one-year fellowship at Yale with Mendel working on physiological chemistry. On her return to Philadelphia in 1898, she took a position as director of the Clinical Laboratory of the Women's Medical College of Pennsylvania, a post that she held until 1906. In addition, in 1904, she agreed to become bacteriologist with the Philadelphia Bureau of Health. Her first major project was to investigate contamination problems with local dairy products. Rather than using court actions against the farmers, she invited them to her laboratory, showed them the high bacterial levels, and demonstrated procedures that would result in safe milk products.[39] The following year, she accepted the additional responsibility of bacteriological chemist at the Bureau of Chemistry, U.S. Department of Agriculture (USDA).

Her work at the Bureau of Chemistry so impressed its director, Harvey W. Wiley, that he wanted her to head the new division of the Bureau which was charged with the implementation of the 1906 Pure Food and Drug Act. Lisa Robinson has recounted the events that followed:

> Pennington initially was hesitant to apply for the position of Chief of the Food Research Laboratory because she did not think that the Civil Service would give such a job to a woman. Wiley encouraged her to at least take the Civil Service examination. When he received the exams from all the applicants, he changed her name, without her knowledge, to read M. E. Pennington. She earned the highest score on the examination and was therefore offered the job. When Pennington wrote to accept the position, the Civil Service realized that M. E. Pennington was not a man, and told Wiley that since there was no precedent for hiring a woman, the position would have to be offered to someone else. Wiley replied that there was equally no precedent for not hiring someone just because she was female. The Civil Service bowed to Wiley's argument and Pennington was allowed to become the director of the new Food Research laboratory.[40]

The work of this laboratory was crucial. Since the early 1890s, the public had begun to look upon cold storage as a means for unscrupulous commercial poultry and egg producers to keep and then later sell unsafe poultry and eggs. It was the task of Pennington's laboratory to determine the temperatures and other conditions under which poultry and eggs could be safely stored without danger of bacterial growth. Her first step was to define the visible signs of decay and devise reliable chemical and bacteriological tests that could be used to identify deterioration.[41] Then she began to study every step in the process, from the slaughter of the chickens through to the handling in the retail establishments, developing strict standards for each step.[42] This work was her greatest personal success. She remained chief of the laboratory until 1919, having watched it grow from four employees to 55, and during that time, she was an official U.S. delegate to international conferences on refrigeration.

After her resignation, she became a director of American Balsa, a New York company that manufactured low-density insulation for refrigeration units. She was frequently asked to help with problems in the perishable food industry and, as a result, in 1922 she set up a private consulting business that she maintained for 30 years. Concurrently with her work in the bacteriological and chemical aspects of food handling, she became involved in the engineering and construction aspects of the problem. Thus Pennington played a major role in the design of the standard railroad refrigerator car[43], and in later years the design of refrigerated warehouses, household and industrial refrigerators, and freezers. She received many awards, including the Notable Service Medal from Herbert Hoover in 1919, and she was elected a fellow of the American Society of Refrigerating Engineers. She was also one of the earliest women members of the ACS, having joined in 1894, and she received their Garvan Medal (the award for excellence by a woman chemist). An enthusiastic gardener in her spare time, she continued to be active in her later years; for example, at the age of 76, she was a U.S. delegate to the Eighth World Poultry Congress in Copenhagen, Denmark. She died at age 80 of a heart attack following a fall in her apartment.

Willey Glover Denis (1879–1929)

Like Pennington, Willey Denis worked for a period with Harvey W. Wiley at the Bureau of Chemistry of the USDA, but her life took a very different direction.[44] Denis was the eldest of three children of an old respected New Orleans family, her father being a banker. Her A.B. degree, obtained in 1899 from the H. Sophie Newcomb College for Young Women of Tulane University, was in modern languages, and it was only during two years at Bryn Mawr (1899–1901) that she developed an interest in chemistry and geology. She was awarded an M.A. from Tulane in 1902, but continued doing unspecified graduate work until 1905 when she entered the doctoral program in chemistry at the University of Chicago. Though her thesis was on the oxidation of aldehydes, ketones, and alcohols, she also did research in physiology, and it was her well-received publication on the comparative effect of calcium and magnesium ions versus potassium and sodium ions on nerve impulse transmissions[45] that focused her career toward biochemistry.

Upon completion of her Ph.D. in 1907, she taught analytical chemistry at Grinnell College for one term and then joined Wiley at the USDA. She left the USDA in 1909, perhaps believing that her advancement possibilities were limited, and entered medical school at Tulane in the same year. She withdrew, reentered in 1910, and again withdrew, dropping out both times because of harassment. Following her second withdrawal in 1910, she became a research assistant in the biochemical laboratory of Otto Folin at the Harvard Medical School. Folin, a clinical biochemist, was a former student of Julius Stieglitz at the University of Chicago, a background that perhaps accounts for his willingness to accept a woman researcher (it is relevant to note that the Harvard Medical School itself was an all-male institution until 1945). Folin was also an acquaintance of Wiley, and it is quite probable that Wiley suggested Denis apply to work with him.

On that occasion, Denis spent only six months at Harvard, though she produced three research papers during that time. She returned to Tulane for the 1910–1911 academic year where she did unpaid research and then spent the summer of 1911 taking courses in physiology at the University of Chicago. She then resumed her work with Folin, first in the biochemical laboratory and later at the Massachusetts General Hospital. Their research work was prolific and significant. Between 1912 and 1913 alone, Denis was sole author of three papers and coauthor of 18 others with Folin. In the 1917–1920 period, some of Denis's work was with yet another pioneering woman biochemist, Ann Stone Minot (1894–1980). Minot was to forge her own career in clinical chemistry at the Vanderbilt University School of Medicine.[46]

Then, in 1920, Denis was appointed as assistant professor in the department of physiology and physiological chemistry of the medical school at Tulane University. This was probably the first appointment of a woman to a major medical school in the United States. In 1922, she was promoted to associate professor and then, in 1925, to full professor and to head of a new and autonomous department of biological chemistry.

Over the years she continued to be a productive researcher; for example, she produced a classic study on blood composition of different classes of fishes, following a summer spent at the Marine Biological Laboratory at Woods Hole.[47] Many of her 99 publications provided major advances in biochemical analytical techniques, her most significant work being on the development of phosphotungstic and phosphomolybdic compounds as colorimetric reagents[48] and the adaption of the Folin-developed urine tests for use in blood analysis.[49] In 1924, Denis was diagnosed with breast cancer, and though treated, the tumor spread. After going blind, she stayed at home but telephoned the department weekly to check on research progress. She died in New Orleans on 9 January 1929, a month short of her fiftieth birthday. Though Denis was accepted as a member of many professional organizations, she never received any recognition for her work.

Rachel Fuller Brown (1898–1980)

Brown was yet another graduate of the University of Chicago. The elder of two children, she was born in Springfield, Massachusetts to George Brown, a real estate and insurance agent, and Annie Fuller.[50] In 1905, the family moved to Webster Groves, Missouri; then, when her parents separated in 1912, her mother moved back to Springfield taking Rachel and her sister with her. During her high school years, history and Greek interested Brown, but she did not enjoy the sciences. Following graduation, she entered Mount Holyoke College to major in history. During her second year, she found that she had to take either physics or chemistry, and she chose chemistry. This was the turning point in her life, for she really loved chemical analysis, and as a result, her A.B. degree became a joint major in chemistry and history. With chemistry now her passion, she was persuaded by Emma Perry Carr (*see* Chapter 9), chemistry professor at Mount Holyoke, to continue her studies at the University of Chicago, where Carr herself had studied chemistry with Stieglitz.

Brown took Carr's advice, completing an M.S. in organic chemistry the following summer. Upon graduation, she taught science at a combined girls' preparatory school and junior college, but after three years she decided that teaching was not her career goal. She spent the summer of 1924 at Harvard and then returned to the University of Chicago to obtain a Ph.D. with Stieglitz as her thesis advisor.

In addition to organic chemistry research, she elected to take courses and do some research in bacteriology. Her project involved the isolation of a pneumococcal-specific polysaccharide, a compound that identified one of the types of bacteria causing pneumonia. By 1926, she had completed the course work and research projects for the Ph.D., but there was a delay in the scheduling of her oral examination. Her savings exhausted, Brown left Chicago and accepted a position at the Division of Laboratories and Research, New York State Department of Health, in Albany, New York. She had applied to the Division after a friend from her days at Mount Holyoke, Lucena K. Robinson, had recommended the work environment there. Among Brown's projects was a continuation of her work on the several pneumonia-causing bacteria. Before the days of broad-spectrum antibiotics, the most successful treatment was the use of antisera, but each serum was only effective against one specific bacterium type. Thus it was essential to have a means of distinguishing the bacteria by a simple test. Brown, with her supervisor Augustus Wadsworth, showed that each type of pneumococcus had a unique polysaccharide coating, and she devised a simple precipitation test to distinguish them.[51]

In 1933, Brown was asked to represent the Division at a scientific meeting in Chicago. Wadsworth and Stieglitz arranged that she could take her oral examination while in the city, and this she did, finally becoming Dr. Brown. Her research continued to be fruitful, but it was 1948 when her life was to be changed forever. It was then that she was introduced to the microbiologist Elizabeth Hazen.

Hazen, 13 years older, had been a bacteriologist, but Wadsworth needed a mycologist in his team of researchers as fungal infections were becoming of increasing concern. Thus in 1944 Hazen, who was working in the New York laboratories of the Division, spent her spare time studying medical mycology at Columbia University. She began a collection of specimens that was in later years to become the basis of her definitive text on pathogenic fungi.[52] She was soon overwhelmed by fungal samples flooding in for identification, but the major problem was that even when a fungus was identified, there was no suitable treatment to combat the infection. As Hazen and her colleague, Albert Schatz, commented: "... no antibiotic agent approaching the efficacy of penicillin and streptomycin against bacterial infections is available in fungus infections either of the superficial or the deep-seated type."[53] To remedy this, Hazen began looking for microorganisms that had antifungal properties. The antibacterial, streptomycin, had been discovered in soils, so she conjectured that soils might contain antifungals as well.

Wadsworth decided that an interdisciplinary approach was necessary for the solution to the problem and that Hazen and Brown were to be the team. As their biographer R. S. Baldwin remarked: "Thus the collaboration began: Hazen the microbiologist—slight, peppery, impatient, intensely active, and infinitely resourceful; Brown the organic chemist—sturdier and a few inches taller, solidly dependable, seemingly imperturbable, and quietly powerful when necessary. It was an odd combination, but one that worked."[54]

Hazen obtained soil samples from far and wide and endeavored to find and grow soil microorganisms that showed antifungal activity. When positive results were obtained, the culture was sent by mail from New York to Albany where Brown would start the painstaking task of solvent extraction of the active component from the myriad of compounds present—and, of course, this was in the days before high-pressure liquid chromatography (HPLC) and other modern separation techniques.[55] Each isolated extract was then sent back to New York for Hazen to test on two important fungi, *Candida albicans* and *Cryptococcus neoformans*. When an extract was active against the fungi, almost invariably it was far too toxic for use in humans.

Amazingly, the most promising source was obtained by Hazen herself. She had visited friends, the Nourses, in Virginia, and while there, she had taken samples of soil from their farm. One organism from these samples produced two antifungal components, one of the components having a very low toxicity to animals. Hazen grew large quantities of this previously unknown organism (later named *Streptomyces noursei*) so that Brown had enough material to extract usable quantities of the nontoxic compound. The compound appeared extremely active against the two chosen fungi and also against at least 14 other fungi. They originally named the compound "fungicidin", but this name had already been used for another compound, so it was renamed "nystatin" after New York State. In October of 1950, the director of the laboratories, Gilbert Dalldorf, felt that a public announcement had to be made. Brown and Hazen had not completed their study on nystatin, but Dalldorf feared that further delay might result in their being overtaken in the search for antifungals. So he scheduled a presentation on nystatin for a regional National Academy of Sciences meeting and then informed Brown and Hazen that they were on the program.

Following the presentation by Brown (Hazen had refused to participate), pharmaceutical companies rushed to follow up on the discovery. Their response came as a great surprise to Brown and Hazen, but the two scientists realized that the development of the drug was far beyond their capabilities. Brown remarked: "If we had an antifungal that was useful, we were in some way responsible to see that it was made available. But we were in no position to carry out the evaluation in humans, no position to even prepare enough material for the tests in the first place. Obviously, it had to be patented before any company would take it on, but how could we make sure they would carry through?"[56]

Brown also had very strong moral convictions. Whenever she had been awarded a scholarship or grant, even though there was no obligation, she always repaid it in later years so the money was available for another needy person. The solution that Dalldorf found was the Research Corporation, a private foundation that supported research by means of grants and, of particular importance, that organized patenting and licensing of discoveries. The Corporation arranged that the pharmaceutical company E. R. Squibb (later part of Olin Mathieson Chemical Company) would develop and sell the product as Mycostatin while the royalties would be divided equally between the Research Corporation and a special fund, the Brown–Hazen Fund. This fund was committed to support scientific investigation in the biological and related sciences at nonprofit scientific and educational institutions. Over the life of the patent (1955–1976), $6.7 million went into the Brown–Hazen Fund.

Though not receiving any financial benefit from their work, Brown and Hazen were frequent visitors to Squibb ensuring that their "baby" was hastened into production. The two continued their collaboration on antifungals, publishing numerous papers, and finding two new active compounds, phalamycin and capacidin, though neither was to become as useful as nystatin. Many awards were given to them, including the Chemical Pioneer Award from the American Institute of Chemists in 1975. They were the first women to receive this award, and the bylaws had to be changed so that Hazen, a nonchemist, could be joint recipient. Although Hazen died the same year, Brown continued to play an active role in the disbursement of grants from the Brown–Hazen Fund. In addition to her continuing interest in antifungals, Brown was a regular abstractor of papers for *Chemical Abstracts* until she died in Albany at age 81.

Dalldorf, in his reminiscences, characterized the "Golden Age" of research at the Division which made the work of Brown and Hazen possible:

> In the beginning we were all amateur scientists. We were teachers or pathologists or biological chemists who were

beset and fascinated by questions and problems that con-
stantly arose in the wards and the morgue.... Curiosity and
the fascination of advancing knowledge ruled. The material rewards were meager and often uncertain.... Our system had other virtues. It provided so well for fledgling scientists who were eager to try their wings and who profited
from these inconspicuous opportunities to venture small
excursions and to learn by trying. It provided a high
degree of freedom and freedom opens the door to intuition and to innovation. All were free to pursue truth in
their own ways, regardless of the consequences.[57]

Gerty Theresa Radnitz Cori (1896–1957)

Gerty Radnitz, the eldest of three daughters, was born in Prague,
then part of the Austro-Hungarian Empire, to Otto Radnitz, a manager of several sugar refineries, and Martha Neustadt.[58] Radnitz was
tutored at home until the age of 10, then sent to a private girls'
school from which she graduated in 1912. She was encouraged by a
maternal uncle, a professor of pediatrics, to attend university, but her
education lacked the courses in Latin and science that were required.
By 1914, she had learned all the missing background knowledge and
she became one of the few women to enter the Medical School of the
German University of Prague.

While at university, Gerty Cori met Carl Ferdinand Cori. They
were both avid experimenters and even as undergraduates published
a joint research paper.[59] Carl Cori was drafted into the Austrian army
during the First World War, and in the meantime, Gerty became
demonstrator in medicine. After their graduation in 1920, they
moved to Vienna and married there. Carl Cori worked at the University of Vienna medical clinic and the Pharmacological Institute, but
Gerty Cori only had a position as assistant at the Karolinen Children's
Hospital. Food was very scarce in the years after the war, and the physicians at the Children's Hospital were provided with dietary supplements. However, the staff of the hospital refused the supplements so
that they could be given to the patients. As a result of this generosity,
Gerty Cori developed symptoms of a vitamin A deficiency.

Unhappy in his post at Vienna, Carl Cori accepted a biochemistry
appointment in 1922 at the State Institute for the Study of Malignant
Disease in Buffalo, New York. Gerty Cori followed six months later
and obtained a position as assistant pathologist at the same institution. Her initial research was on the effect of X-rays on organs in
mice[60] and then on the insulin content of cancer tissue.[61] Three years

later, she was appointed as assistant biochemist and their joint work began, initially on the metabolism of tumors, then later on carbohydrate metabolism. Sharon McGrayne has summarized this unequal working arrangement which was not uncommon:

> The academic countryside was littered with scientific couples studying botany, genetics, chemistry, and other sciences. Professor husbands and their low-ranking, low-paid [or unpaid] wives often worked together for decades...the women were generally low-level instructors, lecturers, or research assistants while their male partners were professors with tenure. A woman had a permanent position only as long as her relationship with the man continued. In case of divorce or disaffection, the woman could be fired.[62]

In fact, Carl Cori turned down several offers at other universities when they refused to offer corresponding positions to Gerty Cori. Making life particularly difficult were the antinepotism rules or laws in place at many institutions. Though well-intentioned, these regulations more often worked to the detriment of the female partner.

Throughout the 1920s, the Coris were interested in how the body sends energy from one place to another. Their work focused on the starch-like polymer, glycogen, which is found in the liver and in muscles. They showed how muscle converts glycogen to glucose, which is used as the energy source. Lactic acid, a product from the use of glucose, is then cycled back to the liver where it is reutilized in the synthesis of more glycogen, which can then be transported back to the muscle as glucose.[63] This is now known as the Cori cycle and once this cycle had been established, much of the remainder of their research was focused on filling in the details. As they delved deeper into the cycle, they discovered a totally new glucose derivative, glucose-1-phosphate, now known as the Cori ester, and one of the intermediates in the conversion of glycogen to glucose.

In 1931, the Coris moved to Washington University in St. Louis, where Carl became chair of the Department of Pharmacology and Gerty Cori was appointed a research fellow at a token salary. Carl was quiet and reserved while Gerty was the opposite. As McGrayne commented: "Smoking incessantly, scattering ashes over the laboratory tables, Gerty had an infectious exuberance. When she read or learned something exciting, she raced down the hall, clack-clacking in her pumps to tell Carl. With her insatiable curiosity, investigative zeal, and tremendous drive, she found research exhilarating."[64] Her devotion to work meant that during her one pregnancy, she worked in the laboratory until the last moment before her son was born in 1936. This passion for her work did not always endear her to people as she had a very sharp tongue for those whose performance did not meet her exacting standard.

In the late 1930s and early 1940s, Gerty Cori finally received long-overdue promotions considering her lengthy publication record. First, in 1938, she was made a research associate and then, in 1942, an associate professor of research. About this time, Gerty Cori was the decisive influence in a change in direction of their research toward enzymology. In fact, of the 10 papers published by the Coris during 1938 and 1939, Gerty Cori was the major contributor to seven. Resulting from this shift in emphasis, they discovered phosphorylase,[65] the enzyme that breaks glycogen down into the Cori ester.

Gerty Cori was finally made a full professor in 1947, the year that she and Carl were co-recipients of the Nobel Prize for Physiology and Medicine. They generously shared part of their monetary award with the several co-workers on the phosphorylase project. The same year, Gerty was diagnosed as having a bone marrow disease, possibly initiated by her X-ray studies during the early years at Buffalo. Though requiring numerous blood transfusions, she continued research work until the end, finally dying of liver failure in 1957. The Coris' research had been truly collaborative but many of the honors that they received were not equal. For example, only Carl Cori was elected to the Royal Society, and while Carl Cori received the Willard Gibbs Medal of the ACS, Gerty Cori had to settle for the ACS Garvan Medal for women chemists.

Gertrude Bell Elion (1918–)

Elion was another woman Nobel laureate in the biochemical field during our time frame. The elder of two children, she was the daughter of Robert Elion and Bertha Cohen.[66] Her father emigrated to the United States from Lithuania at age 12, and her mother emigrated from Russia at age 14. Robert Elion put himself through medical school in New York and become a dentist. Her biographer, McGrayne, described Trudy's (Gertrude's) personality:

> Trudy was a shy bookworm with an insatiable thirst for knowledge. "It didn't matter if it was history, languages, or science, I was just like a sponge." She idolized Louis Pasteur and Marie Curie—"people who discovered things"—

and devoured popular science books like Paul de Kruif's *Microbe Hunters*. "Those books were so exciting", remembered Elion. "It was like reading a novel. It was a mystery story that they could solve, and they became people. They weren't just names." Her heroes had to be discoverers, but their sex did not matter.[67]

Despite her father's bankruptcy following the stock market crash, Elion was encouraged to continue education beyond high school. Elion commented: "Among immigrant Jews, their one way to success was education, and they wanted all their children to be educated. Furthermore, it's a Jewish tradition. The person you admired most was the person with the most education. And particularly because I was the firstborn, and I loved school, and I was good in school, it was obvious that I should go on with my education."[68] In 1933, she entered Hunter College, New York, the women's college of the City University of New York. With her high grades, she could have majored in almost anything, but having seen her grandfather die of stomach cancer, she decided that her long-term goal was cancer research. However, disliking animal dissection, she chose chemistry over biology as a major. Unfortunately, when she graduated in 1937 with the highest honors in chemistry, her dreams did not come true. Though she applied to 15 graduate schools, none would offer her any form of financial assistance. Being rejected despite such high grades suggests that, at least in some cases, she must have been discriminated against on grounds of sex and/or her Jewish religion. Her eyes were opened to such possibilities when she was turned down for a job because, though excellently qualified, she was told she would have been a distracting influence for the all-male staff.

For seven years, Elion worked marginal and temporary jobs, trying to obtain the experience necessary to enter the world of research chemistry, at one point offering to work in a lab for free, just to gain knowledge. After taking education courses, she became a substitute teacher in the New York City high schools, which enabled her to work on a master's degree at New York University at nights and weekends. Finally, with the war in progress, the demand for chemists was such that even women were being hired. First, she worked at Johnson & Johnson, but with the subsequent closure of the plant, she had to look elsewhere. Her father had just received a sample of a painkiller from a company named Burroughs Wellcome, and he suggested she apply there. Elion went for an interview on a Saturday when George Hitchins was working. He had a vacancy and was very impressed by this outspoken and brilliant young chemist, though his (female) research assistant, Elvira Falco, had doubts whether such an elegant woman could really handle the dirty bench work.

Burroughs Wellcome was the ideal place for Elion, and she accepted the position as senior research scientist. The company had been founded in Britain by two American pharmacists as a charitable trust to discover drugs to treat serious, incurable diseases—the same goal as Elion. The working conditions were less than ideal, though. A baby-food plant below the lab caused the summer floor temperature to rise to 60 °C. Despite this, Elion enjoyed the lab environment and became a good friend of Falco, though Falco was more outgoing, enjoying water-bottle fights with the visiting British chemist, Peter Russell. Falco and Elion would attend performances of the Metropolitan Opera together, and Elion also became close to the Hitchins family, often vacationing with them.

Hitchins disliked the traditional hit-and-miss method of drug research, believing that more scientific methods were needed. Though all cells need nucleic acids, it was known that bacteria and tumor cells had a particularly high demand to sustain their rapid growth. This,

Gertrude Elion (1918–) in her laboratory. (Courtesy of Glaxo Wellcome.)

Hitchins argued, was their Achilles heel. He assigned a study of the purine bases to Elion, and he let her follow her instincts. This was a wise move, and Elion was to publish a total of 225 research papers from her studies. It was 1950 when success first appeared with the synthesis of 2,6-diaminopurine and its successful testing against leukemia at the Sloan-Kettering Memorial Hospital. Too toxic for widespread use, Elion searched for less poisonous analogs and discovered 6-mercaptopurine.[69] This was an improvement, but she believed that the only method of finding the best anticancer agents was to understand the metabolism of the drugs. This became one of her major research goals. Another compound that she synthesized, the 1-methyl-4-nitro-5-imidazolyl derivative known as azathioprine (Imuran), proved to be an effective immunosuppressive drug.[70] Yet another, allopurinol, proved to inhibit uric acid synthesis, making it a choice for the treatment of uric acid excess, for example, in gout.[71]

Hitchins and Elion, together with Falco and others, were a very productive team. However, they had very different views of their working relationship: Hitchins (according to Elion) saw himself as the boss, while she considered the interaction as a collaboration of equals. She added: "He [Hitchins] can be very patronizing. He perceives that he started it all.... But actually he was always willing to listen to suggestion. I said I wanted to study metabolism in 1953; nobody was studying drug metabolism. He just let us do what we thought we should be doing."[72]

In 1967, Hitchins retired from active research, and Elion became head of the Department of Experimental Therapy. Now on her own, Elion could show that she had not just been Hitchins's assistant. She heard that one of her early purines had some antiviral activity. This was totally unexpected, for it was generally accepted that a drug treatment against viruses was impossible. Elion sent a sample of a related compound, arabinosyldiaminopurine, to England for testing and was told that it was active against *herpes simplex* and *herpes zoster*. The head of the Organic Chemistry Division at Burroughs Wellcome, Howard Schaeffer, and his assistant, Lilia Beauchamp, took on the task of altering the sidechains on the purines hoping to produce an even more active compound. In this, they were successful, synthesizing acycloguanosine, known as acyclovir (Zovirax). Once its efficacy had been shown, Elion's team went to work, trying to follow its metabolism so that they could understand why it was so nontoxic and specific. By showing that the conversion of acyclovir into a cell-toxic compound required an enzyme uniquely produced by the virus itself, they proved that acyclovir was safe for administration without fear of harmful side effects. In fact, until her work, it was not realized how many enzymes were unique to viruses, providing potential pathways of combating virus replication.[73]

During her early years at Burroughs Wellcome, she had enrolled in the Ph.D. program at Brooklyn Polytechnic Institute, but dropped out after two years when she was told that she would have to attend full-time rather than just evenings as she had done until then. Thus she was particularly pleased to receive honorary doctorates from George Washington University and Brown University. Elion retired in 1983. Her greatest recognition was sharing the Nobel Prize in Medicine in 1988, one of the co-recipients being Hitchins. A year later, her former research unit used her approach to synthesize the anti-AIDS drug, azidothymidine (AZT).

Edith Gertrude Willcock (1879–1953)

Though we have focused on American women biochemists, there are several claimants to fame by the members of F. G. Hopkins's group. Our one representative here is Edith Willcock, whose work has been recognized as a classic in biochemistry.[74] Willcock was born at Albrighton in England[75] and she attended the King Edward VI High School for Girls in Birmingham, where science has been a crucial part of the education from the 1880s.[76] Whereas in the United States it was particular women's colleges, such as Bryn Mawr and Mount Holyoke, that produced many women scientists, in Britain, it was this one high school that had a disproportionately large influence.

Willcock entered Cambridge University in 1900 where she stayed as a student until 1905. During her undergraduate studies, she did research with the biochemist W. B. Hardy on the oxidation of iodoform by radiation from radium. Of more significance, on her own she published a paper on the effects of radiation on animal life, probably one of the first studies that showed the damaging effects of radiation.[77] But she had been inspired by the lectures of Hopkins, recalling that the impression left on her mind was "less a statement of information than a realisation of the existence of vast unexplored tracts and the unfolding of immense opportunities for research."[78] Resulting from this fervor, Willcock became a research fellow with Hopkins, and it was her work with him between 1905 and 1909 for which she should be remembered. At the time, 1906, it was believed that diets were complete as long as they contained the appropriate chemical functional units. In particular, it was known that the indol unit was required for certain biological functions. Willcock and Hopkins[79] were the first to show that diets had to contain specific molecules, in this case, tryptophan, and that other indol-containing compounds would not function in its place. It was this initial study that focused attention on the essentiality in diets of certain amino acids.

Willcock married John Stanley Gardiner, professor of zoology at Cambridge, in 1909 and her research publications ceased. During the First World War, she was a local consultant for the British Ministry of Agriculture on the raising of rabbits and poultry, and she wrote leaflets on the subject for public distribution. Subsequently, she became an advisor on oyster culture. Willcock had several other interests, being a recognized water-color artist, singer, and author of a popular children's book, *We Two and Shamus*, published in 1911. She died on 8 October 1953.

Maud Lenora Menten (1879–1960)

All biochemists and most physical chemists know of the Michaelis–Menten equation, but few realize that Menten was a woman. Maud Menten was born in Port Lambton, Ontario, Canada. She graduated from the University of Toronto, receiving a B.A. in 1904 and an M.B. in 1907.[80] Her M.B. research, on the distribution of chlorides in nerve cells and fibers, was coauthored with her mentor, A. B. McCallum. Menten was appointed fellow at the Rockefeller Institute for Medical Research for the 1907–1908 year where she studied the effect of radium on tumors.[81] She then returned to the University of Toronto where she was awarded an M.D. in 1911.

In 1912, Menten traveled to Europe, joining Leonor Michaelis at the University of Berlin. It was the work there that was to give her eternal recognition—an equation named after her and her colleague, the Michaelis–Menten equation.[82] This equation for enzyme-catalyzed biological reactions relates the rate of formation of product to the concentration of enzyme and of substrate (reactant). Staying just one year in Berlin, Menten traveled to the University of Chicago, where she obtained a Ph.D. in biochemistry in 1916. Unable to find an academic position in her native Canada, she joined the medical school at the University of Pittsburgh. She was appointed assistant professor of pathology in 1923 and was promoted to associate professor in 1925. At the same time, she was clinical pathologist at the Children's Hospital of Pittsburgh. This work was actually three positions: surgical pathologist, postmortem pathologist, and hematologist. It was commented that "[s]he was made aware of every puzzling or interesting case admitted to Children's Hospital during her tenure. The pediatric residents flocked to the laboratory to consult her, and she was never too busy to listen."[83]

Graduation photo of Maud Menten (1879–1960). (Courtesy of University of Toronto Archives.)

Menten maintained an active research program, authoring and coauthoring more than 70 publications. Among her later work, the two most important contributions were the first use of electrophoretic mobility in studying human hemoglobins[84] and the use of an azo-dye coupling reaction for the study of alkaline phosphatase in kidneys.[85] She managed to juggle her clinical, teaching, and research duties, together with her other responsibilities, by working 18-hour days. George Fetterman, who had been one of her students, commented: "She was full of ideas and was highly critical of researchers who ran out of them. With her, it was a matter of 'What have you discovered recently?' In discussing the career of a world-renown[ed] physician who had been awarded the Nobel prize, Dr. Menten's comment to me was 'What has he done since?' Fortunately, she was much more lenient with medical students than with Nobel prize winners."[86]

Though she was obviously devoted to her work, Menten still found time for other pursuits. She was an avid mountain climber. Also, she was very artistic, some of her paintings being exhibited with the Associated Artists of Pittsburgh, and she was very musical. Her prized possession was a model-T Ford that she drove for many years.

Despite her productivity, promotion to full professor did not come until 1949 when she was 70 years old and one year before her retirement from her position at Pittsburgh. However, this was not the end of her active research life. In 1950, after retirement, she returned to Canada to perform cancer research at the Medical Institute of British Columbia. Because of ill health, Menten resigned her position in British Columbia and returned to Ontario where she died on 20 July 1960 at Leamington. In their obituary, Stock and Carpenter remark: "Maud Menten was untiring in her efforts on behalf of sick children. She was an inspiring teacher who stimulated medical students, resident physicians and research associates to their best efforts. She will long be remembered by her associates for her keen mind, for a certain dignity of manner, for unobtrusive modesty, for her wit, and above all, for her enthusiasm for research."[87]

Dorothy Maud Wrinch (1894–1976)

A few women scholars moved from field to field, and Wrinch provides one such example. She was the single child of Ada Minnie Souter and Hugh Edward Hart Wrinch, a mechanical engineer.[88] Though she grew up in southern England, she was born in Rosario, Argentina. From high school, she won a scholarship in 1913 to Girton College, Cambridge University, graduating in 1916 with the highest ranking of the year in mathematics. After obtaining an M.A. from Cambridge in 1918, she accepted a position as lecturer in mathematics at University College, London University, where she remained until 1921. Then, she returned to Girton as a research scholar. Wrinch married John William Nicholson, director of studies in physics and mathematics at Balliol College, Oxford University, and she moved to Oxford in 1923, becoming tutor in mathematics to the five women's colleges of the University. The position was annually renewed until 1927 when it finally became a long-term appointment.

Wrinch was a prolific and versatile scientist. Between 1918 and 1932, she published 16 papers on scientific methodology and the philosophy of science, together with 20 papers on aspects of pure and applied mathematics. Though this versatility can be looked upon as the mark of a true scholar, it also had its negative aspects as Pnina Abir-Am has pointed out:

> Moreover, her prolific output tended to be unfocused because of her splitting her research effort between mathematics and philosophy, a legacy of her youthful infatuation with her teacher and friend Bertrand Russell. Even within mathematics she vacillated between the pure and applied fields, which were separated by conflicting ideologies. This was a legacy of her collaboration with her husband, a mathematical physicist, and her father, an engineer. As a result, Wrinch lacked a strong disciplinary reputation, despite her substantial talents in all of the above mentioned fields. All this was further compounded by the gender problem which limited women's opportunities at Oxbridge [Oxford and Cambridge] and by a class problem.... [Wrinch having come from a middle class environment and now in an upper class milieu].[89]

During this period, she also had one child, Pamela, and her experiences of combining academic duties with marriage and motherhood led to her authoring a book on the subject under a pseudonym.[90] Unfortunately, her life was to be suddenly changed by the permanent institutionalization of her husband as a result of alcoholism in 1930. A major problem was the sudden reduction in income, but it also caused Wrinch to reassess her direction. At this point in time, biological architecture had begun to fascinate her—the application of mathematical topological techniques to the interpretation of biological molecular structures. Wrinch herself commented:

> I had, however, long had a consuming interest in physiology and chemistry, and I had always hoped to find special-

ists in this field with whom I could develop certain ideas. It proved impossible to arrange such a collaboration, since the mathematical point of view was difficult to link up with the point of view of the professional chemist. At this time, then, it became clear to me that I had but two choices, either to abandon the attempt to develop these ideas or to undergo apprenticeship in chemistry sufficiently extensive to enable me to formulate the ideas in a form suitable for development by specialist workers. I chose the latter course and spent a year's leave of absence from Oxford on the continent of Europe beginning an apprenticeship in many different laboratories.[91]

Though her diversity of interests was a problem in some ways, it resulted in her becoming one of the five founder members of the Biotheoretical Gathering. This small elite of British scientists focused on redefining the relationships between scientific disciplines and between science and philosophy. Wrinch's contribution dealt with applications of topology to experimental embryology. In the mid-1930s, Wrinch developed a new theory of protein structure, a theory that combined ideas of mathematical symmetry with the concept of a covalent bond between the CO of one amino acid with the NH of another amino acid. The cyclic structure that resulted from pairs of these bonds, she referred to as a cyclol bond.[92] The series of cross-links proposed by Wrinch would result in sheets that could fold to give closed geometric figures. According to her calculations, some of these structures would contain $72n^2$ amino acids, where n was a small integer. Fortuitously, two researchers found egg albumin to contain 288 or 72×2^2 amino acids.[93] This, and many other pieces of evidence, seemed to point to the correctness of Wrinch's model of proteins, and many scientists, particularly Irving Langmuir, supported her proposal.

At the same time, there were opponents; for example, neither Dorothy Hodgkin nor J. D. Bernal believed in the cyclol hypothesis. Hodgkin commented: "She [Wrinch] was a friend of ours, and as a mathematician was very anxious to become acquainted with biology and biochemistry and to contribute to the problems on which we were working.... We were friends of hers, and had helped to develop her theories, but we did not believe in them, and that was our trouble."[94] Linus Pauling was another critic and the exchanges between Pauling and Wrinch became acrimonious and often personal. Though neither side was untarnished by the dispute, it is undeniable that Wrinch clung to her general cyclol hypothesis of proteins long

Dorothy Wrinch (1894–1976), right, showing her cyclol protein model to Katherine Blodgett (1898–1979), center, and Wanda K. Farr in 1936. (Courtesy of Sophia Smith Collection, Smith College.)

after it had been discredited by X-ray crystallographic evidence.[95] Hodgkin delivered her verdict in the affair as follows:

> Gradually and relentlessly, after the war, chemical and crystallographic work proved that protein molecules did not have the cyclol structure. Yet many of Dorothy Wrinch's geometrical instincts and deductions were, in general terms, correct. The structures of protein molecules are not based on simple parallel arrays of peptide chains, yet the packing of molecules within the insulin unit cell is octahedral in character, there are patterns of close packed units over closed surfaces in many spherical viruses, even the 'cyclol' link itself sometimes exists in the

ergot alkaloids. This pleased her—she found it very hard to give up the cyclol theory as a whole.[96]

By this time, Wrinch and her daughter had moved to the United States, spending the 1940–1941 year in the chemistry department of Johns Hopkins University. With the help of O. C. Glaser of Amherst College, whom she later married, Wrinch was given a simultaneous visiting professorship at Amherst, Smith, and Mount Holyoke Colleges for a year. Then she obtained a research position at Smith where she stayed for 30 years until her retirement in 1971. Though her name is always linked with the cyclol controversy, she maintained a stream of publications covering many fields, giving her a total of 192 in all. In chemistry, she turned her attention toward developing techniques for the interpretation of X-rays of complex crystal structures. This work culminated in the monograph *Fourier Transforms and Structure Factors.*[97] Following retirement, Wrinch moved to Woods Hole, where she had given summer lectures during her years at Smith College. She died there in 1976.

Some More Women Biochemists

In every chapter, it was a tough decision to decide which individuals to exclude from the biographical accounts in order to prevent this book becoming encyclopedic instead of selective. The decision was hardest in the case of biochemistry as there were so many women biochemists. The reason for these large numbers, we have argued, was due to the mentoring roles of Hopkins and Mendel. In addition, we suspect that the overlap of biochemistry with molecular biology made the presence of women more acceptable than in most other fields of chemistry.

It is the diffuseness of the boundaries between biochemistry, biology, and medicine that has added to our dilemma of choice for inclusion in this chapter. Thus among other possible candidates for incorporation were the Americans Helen M. Dyer (1895–)[98] and Pauline Beery Mack (1891–1974).[99] Dyer became well-known as a cancer researcher while Mack contributed to the study of nutrition and was active as a chemical educator. A British contender for inclusion was another of Hopkin's students, Marjorie Stephenson (1885–1948). Stephenson was a leader in the field of bacterial biochemistry[100] and the second woman to have been elected to the Royal Society.[101]

However, biochemistry was no different from other fields in having barriers to women's progress. Though biochemical research lab-

oratories were quite significantly populated by women, recognition
for their contributions was rarely forthcoming, particularly in terms
of status and pay. For example, Dorothy Needham (1896–1987),
another from the Hopkins group and a full-time researcher for 45
years, remarked: "I simply existed on one research grant after
another, devoid of position, rank, or assured emolument. In other
words I belong to the generation for whom it was calmly assumed that
married women would be supported financially by their husbands,
and if they chose to work in the laboratory all day and half the night,
it was their own concern."[102]

Chapter 8

Women in Industrial Chemistry

The Effects of World War I—Biographies: May Sybil Leslie (1887–1937); Florence E. Wall (1893–1988)—The Interwar Period—Biographies: Katharine Burr Blodgett (1898–1979); Kathleen Culhane Lathbury (1900–1993)—Barriers Against Women in Industry

The Effects of World War I

The First World War was to open the doors of industry to women chemists. The 1914–1918 War is sometimes called the Chemists' War as its prosecution demanded ever-increasing quantities of explosives, poison gases, optical glass, synthetic dyes, and pharmaceuticals.[1] However, this point was slow to permeate military establishments, and many chemists were drafted for regular military service, including the brilliant Henry Moseley, who was killed at Gallipoli.[2] As the war progressed and severe shortages of chemicals occurred, more and more women were pressed into chemical-related work.

The vast majority of the women were unskilled, simply working at specific synthesis tasks and following exact recipes to produce the enormous quantities of TNT, nitroglycerine, ammonium nitrate, and ammonium perchlorate that were required by the explosives industry.[3] In fact, the proportion of women in chemical factories was as high as 88%. The wages paid were usually two-thirds of the male rate for the same task, as the argument was commonly made that three women were equivalent to two men. The work was hard, often very dangerous, and it led frequently to debilitating effects from the toxic chemicals. TNT poisoning was among the worst health problems, the

165

sufferers being called "canary girls" as a result of the yellow color of their skin. The medical personnel had orders that only the most seriously affected by chemical poisoning were to be given time off from work; thus many women suffered permanent health damage and some died as a result of their continued exposure to TNT. Yet the experience was not totally negative: most of the women reveled in the camaraderie of the workplace, and for many, the well-balanced, nutritious meals served in the works canteens was much better food than they had ever had before the war.[4]

Little is known about the skilled women chemists who were assigned to war duties. They were obviously much fewer in number though they certainly did exist.[5] For example, when British male analysts were placed on the draft list in the fall of 1916, the University of Sheffield instituted a one-month intensive analytical course for women chemists that provided training in "accurate weighing, filtration, titration, general manipulation, and calculations."[6] At the end of the course, the women graduates were expected to be able to do rapid determination in iron and steel samples of carbon, silicon, manganese, sulfur, and phosphorus. By the end of the war, 96 women had graduated and taken industrial and government positions. One of the replacement analysts was Ada Hitchins, who was Frederick Soddy's research assistant through much of his career.[7] She was assigned in September 1916 to work in the British Admiralty Steel Analysis laboratories for the remainder of the war.[8]

The United States did not enter the war until much later, but long before formal hostilities occurred, the loss of chemical sources from Germany had resulted in a massive expansion of the American chemical industry.[9] To satisfy the demand for industrial chemists, of necessity, the doors were opened to women. In fact, the industrial demand for female analytical chemists between 1916 and 1918 became so great that Mount Holyoke College added analytical chemistry courses to its curriculum to satisfy the need.[10]

Biographies

May Sybil Leslie (1887–1937)

The only woman chemist to receive recognition for her war work (as far as can be established) was May Sybil Leslie. Leslie was born in Yorkshire, England, and she studied chemistry at the University of Leeds.[11] She graduated with first-class honors in 1908. The following year, she was awarded an M.Sc. for research with H. M. Dawson on the kinetics of the iodination of acetone, work that has since become a classic in its field.[12] In that same year, 1909, Leslie was awarded a scholarship which she decided to use to work with Marie Curie in Paris. Her letters from Paris to Arthur Smithalls, Professor of Chemistry at Leeds, are among the few accounts of life in the early Curie laboratory.[13]

Leslie's work revolved around the extraction of new elements from thorium. For a chemist used to working with grams of pure chemicals in beakers, the manipulation of kilogram quantities of minerals in huge jars and earthenware bowls must have been a completely new experience. Leslie spent two years (1909–1911) with Curie, after which she took a position with Ernest Rutherford at the Physical Laboratory of the Victoria University, Manchester. There she continued with her work on thorium and extended her studies to actinium during the 1911–1912 year.

After leaving Manchester, she spent two years as a science teacher at the Municipal High School for Girls in West Hartlepool. During this time, she managed to resume research with Dawson; this work was on ionization in nonaqueous solvents. From 1914 to 1915, Leslie held a position as Assistant Lecturer and Demonstrator in Chemistry at the University College in Bangor, Wales. Then in 1915, she entered the world of industrial chemistry, being hired to work at His Majesty's Factory in Litherland, Liverpool, a position that she obtained as a result of the call-up for military duty of the male research chemists (it is possible that the position at Bangor was offered her for the same reason). Her initial rank was that of Research Chemist, but in 1916, she was promoted to Chemist in Charge of Laboratory, a very high position for a woman at that time. Her research involved the elucidation of the pathway in the formation of nitric acid, and the determination of the optimum industrial conditions for the process.[14] This work was vital for the munitions industry, which required massive quantities of nitric acid for explosives production.

One of her young assistants at Litherland, Edward Rogans, recalled that Leslie had her own office attached to the laboratory, and a young woman graduate, Miss Dicks, as her assistant whose task was to collect details of all work performed in the laboratory by the six staff.[15] The laboratory was responsible for the analysis of the different chemicals used in the production of explosives, but the most dangerous work was the collection of nitric acid samples from the nitric acid absorption towers. In June 1917 the Litherland factory closed[16], and Leslie was transferred with the same rank to the H. M. Factory in Penrhyndeudraeth, North Wales. Leslie was awarded the D.Sc. degree in

1918 by the University of Leeds, mainly in recognition of her contribution to the war effort (the D.Sc. degree is awarded for exceptional merit, and it is not the same as a Ph.D.). The referee of her application for the degree reported on her work with Dawson, Curie, and Rutherford and then described her industrial research:

> The remaining set (of work) comprises two joint papers[17] and four independent inquiries and were carried out at H. M. Factory at Litherland. The confidential nature of these investigations precludes any reference to the subject matter in this report; but they appear to me to constitute the most weighty claim in Miss Leslie's application for the doctorate. The problems she has had to solve are not only of the first importance at the present time, but have been attacked in a manner showing unusual resource as well as novel methods of procedure.[18]

With the return of the surviving male chemists at the end of the First World War, Leslie lost her government position. She returned to the University of Leeds as Demonstrator in the Department of Chemistry in 1920, being promoted in the following year to Assistant Lecturer. Leslie then moved to the Department of Physical Chemistry in 1924 and was promoted to a Lecturer in 1928 (this rank is somewhat equivalent to Associate Professor in the North American system). In 1923, Leslie had married Alfred Hamilton Burr, a Lecturer in Chemistry at the Royal Technical College, Salford.[19] Burr had also worked at the H. M. Factory in Litherland in 1916 and presumably they first met there. She continued to be an active researcher at Leeds after marriage, and as well, the famous British chemist J. Newton Friend invited her to author one volume of the classic series *A Textbook of Inorganic Chemistry* and to coauthor another.[20]

According to University records, Leslie resigned her position at the University in 1929. It is unknown whether her resignation was due to the difficulty of working so far from her spouse, or for health reasons. In 1931, when Burr was appointed head of the Chemistry Department at Coatbridge Technical College, Scotland, she moved with him. After Burr died in 1933, Leslie moved back to Leeds, resuming her research work at the University. Her first project was the completion of her deceased husband's research on wool dyes, then she returned to her interest in the mechanisms of reactions. In addition to performing research work, she was employed as subwarden of a women's hall of residence (Weetwood Hall) at the University from 1935 to 1937. Margaret Rossiter has commented that being sidetracked to Dean of Women, or other women-oriented positions, was a common fate for women scientists.[21]

Leslie died at Bardsey, near Leeds, on 3 July 1937, having given up research only a month earlier. No cause of death was recorded, but it was quite possibly radiation-related considering her exposure to high levels of radioactivity during her research work in Paris. Leslie's chemical abilities were certainly well-accepted during her life; for example, she had been elected as Associate of the Institute of Chemistry in 1918 and Fellow of the Chemical Society in 1920.[22] As well, Newton Friend must have thought highly of her abilities to have made her a contributing author. Her obituary in the *Yorkshire Post* noted that Leslie was "one of the University's most distinguished women graduates."[23] Her former supervisor, H. M. Dawson, commented that her research reputation was "deservedly high" and that as a teacher she was "exceptionally gifted." He concluded: "To her intimate friends she was known as a woman of the highest ideals, of wide human sympathies and of great earnestness of purpose. Her reticence and innate modesty limited the circle of her acquaintances, but such restriction would doubtless count for very little in comparison with the respect and sincere regard of those who were privileged to enjoy her confidence."[24]

Florence E. Wall (1893–1988)

An industrial pioneer on this side of the Atlantic was Florence Emeline Wall. Wall was born in Paterson, New Jersey, the descendant of Irish settlers.[25] She attended both the Academy and College of Saint Elizabeth, Convent Station, New Jersey, where she obtained a B.A. in chemistry with a minor in English, together with a diploma in education in 1913. Her initial career was that of high school science teaching, but in 1917, she responded to the "call" for women chemists and took the first of many positions in the U.S. chemical industry.

Her first work was with the Radium Luminous Material Corporation of Orange, New Jersey. This company, like others in the field, employed a large number of women to paint a radium solution on the numerals of watch dials, the radioactive decay of the radium causing the numbers to glow in the dark (of particular benefit to night warfare). This activity was obviously hazardous in itself, but to make it a virtual death sentence, workers habitually licked the radium-coated paintbrush tip to provide a better point for painting. Over the next few years, many of the women developed lip and mouth cancers, death usually following soon after. A letter to Marie Curie from Florence Pfaltzgraph, an American journalist, described the sufferings of the women radium painters.[26] Pfaltzgraph asked Curie whether she had "discovered anything which might benefit these women." Curie expressed her sympathies and advised the women to eat calves' liver. Fortunately for Wall, she was assigned to the somewhat less hazardous task of chemical analysis of incoming ores and, later, to measuring the concentration of the extracted radium using an electroscope.

After six month's employment there, she became exhausted from working until 1 or 2 A.M. each day so she joined the Seydel Manufacturing Company to work on benzoic acid derivatives. While she was distilling benzyl acetate, a visitor passed through the laboratory and asked for a sample. About a week later, Wall received an order to appear at the office of the Fellows Medical Manufacturing Company, from where the visitor had come. Here she was informed by a representative of the U.S. Chemical Warfare Service that her sample of benzyl acetate had been analyzed and had been found to be very pure. Apparently, the Fellows Company needed a chemist to run the large-scale production of benzyl acetate and benzyl benzoate, chemicals that were required as solvents for the lacquers to coat the fabric skins of the military aircraft. On the basis of her sample, she was told that it was her patriotic duty to undertake this task. This she agreed to do and she was put in charge of the newly constructed plant.

With the end of the war, the plant was closed and she obtained a position with the Ricketts Laboratory where she worked for a while on mineral analyses, organic analysis, and some forensic chemistry. In 1919, Wall was hired by the U.S. Motor Fuel Corporation to investigate a novel catalyst that was claimed to increase the yield of gasoline during petroleum cracking. After initial evidence that there was a slight improvement in yield, a pilot plant was built. One particular morning, Wall noticed a strong odor of gasoline from the vicinity of the pilot plant, and the subsequent yield was found to be suspiciously high. Confronted, the inventor of the catalyst admitted "spiking" the still to encourage further development of his claim. Wall recommended that the investigation be terminated and, as a result, lost her job.

Unable to find another chemistry position, Wall went to Havana, Cuba, where she did general tutoring. Upon her return to the United States in 1924, she was offered a post in cosmetic chemistry (possibly the offer to a woman chemist was a result of the female-oriented nature of the industry). She accepted the job and this proved to be a fortunate choice, for she was to prosper in this field for much of her working life. From 1929, she was a consultant cosmetic chemist, some of her articles on cosmetics appearing in the *Journal of Chemical Education*,[27] and from 1936 to 1943, she lectured on cosmetics at New York University, receiving an M.A. there in 1938. Wall authored two classic texts in the field,[28] and she became such an authority on the chemistry of cosmetics that she was the first woman to receive the medal of the Society of Cosmetic Chemists. In addition, she was inducted into the Cosmetology Hall of Fame in 1965.[29] Wall continued to be active in chemistry, authoring a review on chemicals for hair treatment when in her 80s.[30] Having been one of the first women elected to Fellowship in the American Institute of Chemists in 1923, she became its honorary archivist at age 93. She died in Fairfield, Connecticut, on 2 October 1988.

The Interwar Period

The First World War had dramatically increased the employment opportunities for women, and in particular, it had opened the avenue of industrial careers at all levels. Yet the end of the war brought with it the termination of these advances. Most of the chemical factories supplying the arms industry closed down immediately when the war was over, and for those that remained, it was expected by almost all males and some females that women would return to their traditional and domestic roles in society. For example, the Illinois Steel Company announced in 1919 that "[t]he women chemists of the Illinois Steel Company not only made good as chemists but showed their fine spirit by resigning in order to make places for the men returning from war work."[31] The company did not elaborate on whether the resignations were given with enthusiasm or under coercion.

A detailed survey of the employment prospects for women scientists in industry during the 1930s shows the depths of the problem.[32] In the first place, during the Depression, the supply of male graduates was sufficient in most areas; hence, the argument was made that there was little need for women scientists. Though some companies automatically rejected any women applicants, a few companies did hire them, though solely for routine testing and analytical work. For these tasks, women were believed to excel due to their "manual dexterity, their delicacy of touch, their conscientiousness and their willingness to bear with a routine under which most men become impatient."[33] There was one other advantage of hiring women—that the salaries offered were less, usually about 80% of those for men.[34]

There were common lines of argument for not hiring women. A major complaint was that, after the substantial investment in training, women had the tendency to marry and leave the position. A second reason was that in industry, researchers were often interchanged between the central research laboratory and the production facilities, where the conditions were considered quite unsuitable for the presence of a woman. However, the study noted that this reason was given by companies that had in the past never rotated its researchers in this way.

Finally, many male scientists considered that women did not have the "research type of mind". This argument considered that men had an enthusiasm for their work which involved them thinking about it through their leisure time as well as their working time. Conversely, it was believed that all women "switched off" as soon as they left the workplace, instead thinking "womanly thoughts." As a result, the flashes of inspiration of the true research scientist were unlikely to occur to women.

Though the atmosphere for women in industry had become quite hostile, there were some success stories and two of these follow: Katharine Blodgett and Kathleen Lathbury.

Biographies
Katharine Burr Blodgett (1898–1979)

Though we know of the professional contributions of Katharine Blodgett to chemistry, we know little of her life and experiences.[35] Blodgett was born in Schenectady, New York, where her father, George Blodgett, who died before she was born, had been a patent attorney for the General Electric Company. Her mother, Katherine Burr, took her and her older brother to France and Germany so that they could learn the languages while they were still young. The family returned to New York about 1906, and Blodgett completed her school career at the small private Rayson School in New York City. Upon graduation in 1913, Blodgett won a scholarship to Bryn Mawr College, where her physics professor, James Barnes, persuaded her to become a research scientist.

During her senior year at Bryn Mawr, a former colleague of her father, the physical chemist Irving Langmuir, gave her a tour of the General Electric (GE) Research Laboratories. He suggested that she might obtain a research position at GE provided that she complete additional science courses. To accomplish the course requirement, she took an M.S. at the University of Chicago, which she completed in 1918 and, as a result, joined the GE Research Laboratories to become their first woman scientist.

Initially, and for many years, she worked with Langmuir (a friend of Dorothy Wrinch; *see* Chapter 7). Langmuir had established the properties of monomolecular films on water.[36] In 1919, he reported to the Faraday Society the transfer of such floating monolayers to solid surfaces, noting that "the writer is much indebted to Miss Katharine Blodgett for carrying out most of the experimental work."[37] Blodgett worked with him on a variety of scientific problems over the following years. She left GE temporarily from 1926 to 1928, spending those years at the Cavendish Laboratory of Cambridge University where she worked as a research student with Ernest Rutherford, devising a method of measuring the mean free path of electrons in ionized mercury vapor. This work gained her a Ph.D. from Cambridge, the first woman to receive a doctorate in physics at that university.

In 1933, Langmuir and Blodgett returned to their studies of thin films, hoping to use them as lubricants in jeweled bearings in meters. As part of this research, she devised ways of making films of calcium stearate that were exact numbers of molecules thick. This work proved to be of tremendous importance, and these films are now known as Langmuir–Blodgett multilayers.[38] Blodgett commented: "You keep barking up so many wrong trees in research. It seems sometimes as if you're going to spend your whole life barking up wrong trees. And I think there is an element of luck if you happen to bark up the right one. This time I eventually happened to bark up one that held what I was looking for."[39]

She also showed that reflection from glass surfaces could be minimized by adjusting the thickness and refractive index of the coating, which led to the development of nonreflecting glass.[40] It was the application of thin film technology to make nonreflective glass which made her name. To illustrate the regard in which she is held, a special 1980 issue of the journal *Thin Solid Films* was dedicated to Blodgett, and in it, all the research papers describe work that followed from her pioneering studies. Though she is renowned for the thin film research, she worked also in other areas. For example, during the Second World War, she developed ways to prevent icing of aircraft wings, and she helped produce a more effective method of generating smoke screens. She received several awards for her work, including the Garvan Medal, the ACS award for excellence for American women chemists.[41]

Blodgett remained in Schenectady for almost all of her life. One of her colleagues at GE commented: "She was a modest, unassuming

Katharine Blodgett (1898–1979). (Courtesy of Hall of History Foundation, Schenectady, New York.)

person, small of stature but with a quick wit and a twinkling eye.... The methods she developed have become classic tools of the science and technology of surfaces and thin films. She will be long—and rightly— hailed for the simplicity and elegance of her techniques and for the definitive way in which she presented them to the world."[42] Given this commendation, it is distressing that a review of the first 75 years of research work at General Electric makes no mention of her name.[43]

Blodgett was a very private person; hence we know nothing of her thoughts and personal experiences at GE. A lifelong Presbyterian, outside of her work she enjoyed cooking, bridge, and outdoor pursuits, spending much of her free time at her retreat at Lake George, New York.[44] She died in Schenectady on 12 October 1979. Like Pockels, Blodgett's name has been forever linked with her work. Very thin films deposited on solid supporting surfaces, such as glass or metal, are now known as Langmuir–Blodgett films.[45]

Kathleen Culhane Lathbury (1900–1993)

To appreciate the challenges for a woman in industrial chemistry during the interwar years we have the experiences of Kathleen Culhane Lathbury.[46] British-born Kathleen Culhane was the fourth of six children, her mother dying when she was quite young. Her father insisted on equal opportunity for his daughters and his sons, and he became Lathbury's idol. She attended a girls' high school at which the only science was botany. Admitted in 1918 to the Royal Holloway College, a women's college of the University of London,[47] she discovered that chemistry was her real interest, and she graduated in 1922 with an honors degree in chemistry. Wanting to enter the chemical industry, she was extremely frustrated that employers would not take an attractive young woman seriously for chemist positions. In fact, she was only considered for interviews when she signed her applications "K. Culhane" rather than "Kathleen Culhane." However, once her gender became apparent at the interviews, she failed to obtain any of the positions (the ruse was more successful for her chemist daughter in the 1950s). Marriage was scarcely a survival option for there were few single males of her generation who had not been killed in the First World War. Finally, she obtained work as a school teacher and later, a private tutor.

Joining the Institute of Chemistry proved to be the turning point in her fortunes. Through the Institute, she met Dr. Marrack of the Hale Clinical Laboratory of the London Hospital. Marrack allowed her to gain experience of medicinal chemistry by permitting her to do emergency blood sugar determinations in her free time and without pay. After two years of combining teaching with unpaid analytical work, Lathbury obtained an industrial chemistry position with a manufacturer of lacquers and enamels. However, her delight was diminished after being told by the company that the only reason for hiring her was that they could not afford the salary of a male chemist. The completion of a study of enamel coatings for lightbulbs coincided with an offer of a job back at the clinic, this time as a paid chemical advisor and insulin tester. Not long after, she accepted a position in the physiology department of the large chemical and pharmaceutical company, British Drug Houses (BDH). However, as time progressed, her initial enthusiasm waned:

> I gradually discovered that it was not the intention to employ me as a chemist but as a woman chemist.... I was expected to do all the boring, routine jobs...while anything interesting was handed to one of the men.... The routine work increased enormously in quantity and I took pride in perfecting my technique...thinking it must surely win pro-motion that way. This did not materialize so, by superhu-man efforts and late work, I got some research done which was successful and I was allowed to publish it[48].... The prob-lems I worked on were of my own finding.... I managed to avoid being disliked and was merely regarded as eccen-tric.[49]

As well, the senior staff lunch room was male-only, so Lathbury had to eat her lunch with the women cleaners and clerks. At the time, there were no chemical tests for insulin, and Lathbury significantly improved the physiological testing procedures. The test involved injecting rabbits with an insulin sample and measuring the change in blood sugar reading, and she showed how more reproducible results could be obtained.[50] Her working hours were made more bearable by running the laboratory during the frequent illness of her supervisor, whom she regarded as incompetent. She discovered later that the company directors were well aware of her supervisor's severe short-comings, but they did not act because most of the time, the laboratory was in her capable hands (though of course she did not receive any recognition for this). Her research was recognized outside however, and the Society of Chemical Industry awarded her a grant in 1928 as a "promising young chemist" to travel and present her work in the United States and Canada.

Lathbury had become sufficiently well-recognized in the bio-chemical community to be invited to join the League of Nations Health Organization committee. At the time, there were two forms of insulin known, amorphous and crystalline, and it was important to establish whether the crystals were pure insulin. Four researchers were chosen by the committee, Lathbury being one (though against the wishes of some committee members), to independently deter-mine the physiological activity of the samples. Her results had significant differences from the other three, and the committee concluded that hers must have been in error. She was asked to withdraw them but this she refused to do, convinced of their correctness. Only later was it shown that her results were indeed the more accurate.

Lathbury commenced a study of vitamins in 1933, though she continued with her work on insulin as well. As part of her diverse research studies, she gave a presentation on the need to standardize products containing added vitamins, arguing particularly for enhanced levels of vitamins in margarine. A British newspaper reported on the meeting, describing her as "Miss Kitty Culhane, the Girl Pied Piper of Science" as she had mentioned the use of mice in the vitamin research, and the newspaper added how an "abstruse lecture on vitamins" was delivered by "a pretty girl with blue eyes and bobbed hair."[51]

That same year (1933), she married Major G. P. Lathbury. Having mentioned this to her supervisor, Lathbury was amazed that the directors had to give special approval for the employment of a woman after marriage. The approval was granted in her case because of the importance of her work. In fact, this ban on the employment of married women was not unusual, for it was typical of the backlash of the interwar period in Europe. For example, in 1918, the British Civil Service had instituted a bar on the employment of married women and, during the 1920s, many authorities had banned the employment of married women teachers in grade schools and fired those then employed[52] (such a ban did not, of course, extend to married men teachers).

During the 1930s, Lathbury became a world-accepted expert on the biological assays for vitamins, being awarded Fellowship of the Institute of Chemistry in 1935 for her research contributions to biochemistry. She resigned her position as senior chemist in the same year because of pregnancy. Her inexperienced male successor was offered a starting salary higher than what she had been earning at the time of resignation.

With the arrival of war in 1939, Lathbury offered her services to the war effort, having sent her small daughter to the country. As she later wrote, "Although it was often publicly stated that industry was short of scientists the Appointments Board were unable to tell me of a single opening for which a woman would be considered."[53] Persistence finally resulted in a position as an assistant wages clerk where the senior clerk offered to teach her percentage calculations! After more badgering of officials, she was appointed manager of a statistical quality control department in a munitions factory—at less than half the salary paid to males in the same position. This lower salary had a secondary effect, because salaries determined travel status. Hence, on train journeys to London, her male colleagues traveled first class, while she had to sit alone in a third-class compartment. Amazingly, she was not embittered by her experiences, instead regarding them as a source of amusement.

After the war, she retired from science and took up a second career as an artist, producing paintings that were included in professional exhibitions. It was of great joy to her that her daughter and one of her grandsons became graduate chemists. She died on 9 May 1993.

Barriers Against Women in Industry

Though the First World War opened the doors of industry to women chemists, the doors closed again soon after. For Leslie, her brief time in industry seemed to have been a positive experience and she received recognition of her efforts. Wall was to obtain acclaim in cosmetics chemistry, but one wonders if she would have been quite as fortunate in a traditional chemical industry. Blodgett's success seems to have depended significantly upon support from her mentor, Langmuir, though as we saw above, she was totally ignored in the history of the GE laboratories. For Lathbury, there was one hurdle after another, and in spite of her contributions to the League of Nations and election to fellowships of the Royal Institute of Chemistry and the Royal Statistical Society, little recognition came from the industrial sector. In the chemical industry between the World Wars, women seemed, in general, to have been confined to tasks such as routine analysis. This was often deliberate, for there seemed to have been an obsession among companies concerning departure of women chemists as a result of marriage. Industry managers were convinced that the years and cost of training would be wasted on women chemists. In 1939, W. S. Landis of the American Cyanamid Company commented:

> I have some very interesting examples of the critical situation brought about in a number of cases by sudden resignation through this cause [marriage]. It does place an unfortunate handicap on women and is at the basis of much prejudice to their employment in positions which normally lead to promotion into the higher office. About the only solution that industry has is to retain the woman employee in a more or less routine character of service until she is well across the peak of the age–marriage rate curve.[54]

Margaret Rossiter has shown that the post of chemical librarian was one industrial niche for which women chemists were actively sought.[55] As early as 1910, women were recruited by industry to fill this role, the acceptability of hiring women for the task deriving from the overwhelming dominance of women in the field of librarianship. She notes that even today, chemical information is one area of chemistry dominated by women.

Technical editing was another field thought suitable for women.[56] In 1955, Ethaline Cortelyou urged each graduating woman chemist to consider this field because "She knows what happened to women chemists during the depression and after both World War II and the Korean episode. She has heard the complaint that a woman chemist must be twice as good as a man to get 80 per cent [sic] of his salary."[56] Cortelyou also argued the positive points for a young woman graduate: that technical editing was a suitable bridge during the three or four years between graduation and her first baby; that the career could easily be resumed after her family was older or in the event of a financial disaster; and that there was the companionship of female co-workers in the event of "what most young girls consider the ultimate disaster—a life of single blessedness."[56]

It is no wonder, then, that we find little evidence of successful women chemists in industry until Stephanie Louise Kwolek (1923–)[57] and Mary Lowe Good (1931–)[58], both of whom made their contributions to industrial chemistry much later than the time frame of this book.

Chapter 9

Women in Analytical Chemistry, Chemical Education, and the History of Chemistry

Overview—Biographies: Erika Cremer (1900–1996); Christina C. Miller (1899–); Emma Perry Carr (1880–1972); Mary Fieser (1909–); Hélène Metzger (1889–1944)

Overview

As we discussed earlier, the women chemists of the early twentieth century were found largely in three fields: crystallography, radioactivity, and biochemistry. This does not mean that there were no women in other branches of chemistry—there were—but the individuals were more scattered. In the previous chapter we looked at four women who entered the chemical industry; in this chapter, we will study five women who made significant contributions to other spheres: analytical chemistry, chemical education, and the history of chemistry. As we mentioned in the previous chapter, analytical chemistry became a common field for women primarily as a result of the First World War. Unfortunately, little is known about the individuals who staffed those wartime laboratories. For this reason, we will look at the life and work of two later women analytical chemists, Erika Cremer and Christina C. Miller.

Chemical education has always been a major avenue of employment for women chemists. Among the many notable pioneers in this field was Charlotte Fitch Roberts, the first woman doctorate in chemistry from Yale[1] who, among other accomplishments, wrote a textbook on stereochemistry in 1896.[2] There are very many women chemical educators that we could have chosen for inclusion, but we decided upon Emma Perry Carr and Mary Fieser.

It is most appropriate for a compilation on the history of women in chemistry that we close these biographical chapters with an outstanding historian of chemistry, Hélène Metzger. Metzger was a pioneer in the overthrow of the "great men" approach to the study of the history of chemistry, replacing it with a study of themes and philosophies.

Biographies

Erika Cremer (1900–1996)

Cremer was born on 20 May 1900 in Munich, Germany,[3] the middle child and only daughter of Max Cremer, the inventor of the glass electrode, and Elsbeth Rosmund. Science was virtually in the family genes, for her great-grandfather and grandfather were both scientists and university professors, and her two brothers were also to become scientists. She graduated from high school in 1921 and entered the University of Berlin where she attended lectures given by many of the great scientists of the time, including Fritz Haber, Walther Nernst, Max Planck, and Albert Einstein. Six years later, she received a Ph.D. on the kinetics of the hydrogen–chlorine reaction. Her conclusion, that it was a chain reaction, was so original at the time that she was permitted to publish her results under her name alone.[4]

As a result of her work, she was invited to join Nikolai Semenov in Leningrad to work on kinetics, but instead she chose to stay in Germany and work at the Kaiser Wilhelm Institut für Physikalische Chemie, Berlin, on quantum theoretical problems of photochemistry. Then she worked briefly with Georg von Hevesy at the University of Freiburg on the decomposition of alcohols using oxide catalysts, before returning to Berlin. This time, she worked with Michael Polanyi at the Institut für Chemie, studying the conversion from one spin state of hydrogen, ortho-hydrogen, to the other, para-hydrogen. Unfortunately, in 1933 Hitler dissolved the Institut, as it had been an anti-Nazi bastion. For the next four years, Cremer was unable to find a position. Finally, she joined Otto Hahn's group at the Kaiser Wilhelm Institut für Chemie at Berlin-Dahlem though she moved soon after to the Institut für Physik, where she worked on isotope separation.

In 1938, she received the *Habilitation* qualification from the University of Berlin, which would have entitled her to apply for faculty positions—except for the fact she was a woman. The Nazi German government had passed a law, the Law on the Legal Position of Female Public Servants, which required the dismissal of females upon marriage and which barred all women from senior posts, such as professorships. The hostility toward women intellectuals reflected two concerns of Nazi administrators at the time: that women were usurping men's rights to high-paying jobs and thereby increasing unemployment among men, and that the birth rate was rapidly dropping, thus threatening the future of the German Race.[5]

The start of the Second World War resulted in the drafting of most male chemists, and the resulting shortage, as in the First World War, opened up opportunities for women again. Cremer was able to obtain a position in 1940 as a *dozent* (assistant) at the University of Innsbruck, Austria (Austria had by then been incorporated into the German Reich). She was particularly pleased that the appointment was at Innsbruck, for it enabled her to pursue her hobby of mountain climbing.

It was in Innsbruck that Cremer constructed the first modern gas chromatograph. The selective adsorption of vapors had first been studied prior to the First World War, and during the war, the same principles were used in the gas mask. In such a short review, it is impossible to give credit to all those who made pioneering steps in the development of the gas chromatograph, but it is important to mention the work of Dr. Gerhard Hess at the University of Marburg/Lahn, who used gas adsorption to separate volatile organic acids.[6]

The groundwork was thus laid for Cremer to develop the tool that is now so commonplace. During her studies of the catalytic hydrogenation of ethyne (acetylene), she needed a rapid means of quantitative analysis of ethyne/ethene mixtures. Liquid chromatography was well-known, and she thought it possible to devise a parallel process for the gas phase. With her student, Fritz Prior, she designed and built the first gas chromatograph.[7] This machine used a carrier gas, and to detect the separated components, a thermal conductivity detector. In November 1944, Cremer sent a short paper to the journal *Naturwissenschaften* outlining her work and noting that full details would follow. Unfortunately, with the war reaching its climax, the journal issue containing the account was never published.[8]

In 1945, at the end of the war, Cremer, a German citizen in once more independent Austria, was initially restricted to her home, having to make clandestine visits to the university in a delivery truck. Fortunately, Prior, an Austrian, was able to continue with the research. Cremer was permitted to return to work later in 1945, being appointed director of the Physical Chemistry Institute at Innsbruck, and, in 1951, being awarded the rank of professor. From 1947 to 1950, Cremer, with another assistant, Roland Müller, continued to improve the chromatographic method. Three papers of Cremer on the gas chromatograph appeared in 1951,[9] but these, too, aroused little interest. In 1951 the British scientists A. T. James and A. J. P. Martin, and in 1952 the Czech scientist J. Janak, announced the invention of gas chromatography to the world. Each of their reports was completely ignorant of Cremer's earlier work.

Subsequent to the growth of interest in the technique, Cremer and her students introduced the term "relative retention time", and they devised the method for calculating peak area by multiplying peak height by peak width at half height. As well, they invented head space analysis, the analysis of vapor mixtures above liquids. Cremer continued to work at Innsbruck until her retirement in 1971. Several awards were presented to her for contributions to chromatography, including the first M. S. Tswett Chromatography Award in 1974.[10] In 1990, a special international symposium on chromatography was held in Innsbruck to honor her ninetieth birthday.[11] Cremer died in 1996.

Christina C. Miller (1899–)

Miller was born in Coatbridge, Scotland, on 29 August 1899, the elder of two sisters.[12] In grade school, Miller developed an interest and aptitude for mathematics. At the time, she was told that, as a woman, she could only use her mathematical aptitude for a career as a school teacher. Unfortunately, she had become deaf as a result of childhood measles and rubella, and this was considered a major impediment to a school teaching career. She read an article in a magazine that mentioned the employment potential for women as analytical chemists, and this possibility determined her future.

She won an entrance scholarship to the University of Edinburgh, but to give her a better training for an industrial position, Miller combined the three-year degree program from Edinburgh with a four-year industrial chemistry diploma program from Heriot-Watt College. At the beginning of her graduating year, she talked with the chemist Sir James Walker about her plans for the future. Walker advised her to learn German, then a prerequisite for research because of the importance of the German chemical literature. Thus in her final year, in addition to taking classes, acting as laboratory demonstrator, and performing a research project on organic compounds of arsenic and mercury, she also taught herself German. Despite her hearing difficulty (and her workload), she won the class medal in the university advanced chemistry course, graduated with special distinction in chemistry, and gained a scholarship which allowed her to pursue a higher degree.

Walker accepted Miller as a graduate student, and from 1921 to the completion of her Ph.D. in 1924, she worked with him on the process of diffusion in solution.[13] She was then awarded a two-year Carnegie research fellowship to undertake independent research. The position enabled her to study a problem that had long fascinated chemists, the glow produced when tetraphosphorus hexaoxide oxidized. During this time, she applied for a lectureship at Bedford College, a women's college of London University, but was rejected on the grounds of her deafness. Walker advised her to accept a post as an assistant in the chemistry department at the University of Edinburgh. The position involved the supervision of undergraduate students, but she was allowed to continue with research in her spare time.

By 1929, her thorough research on the oxides of phosphorus showed that traces of elemental phosphorus caused the glow, not the oxides themselves. As a result of her five publications on phosphorus oxides, Miller was awarded a D.Sc. degree, the prestigious Keith Prize by the Royal Society of Edinburgh, and a lectureship with tenure. Shortly afterward, during her research, a glass bulb exploded, the glass fragments blinding her in one eye. This episode persuaded her to change research directions, and she developed an interest in micro- and semimicroanalysis. In 1933, Miller was appointed director of the inorganic laboratory. During the next 28 years, she became renowned for her innovative undergraduate microscale analytical techniques. She was always looking to refine and improve analytical methods, and her many research papers were devoted to this topic, such as the definitive work on 8-hydroxyquinoline as a reagent for magnesium.[14]

Sadly, ill health in the form of otosclerosis, together with family commitments (presumably care of a relative), caused Miller to take early retirement in 1961. Never one to remain idle, she pursued interests in genealogy and in the history of Edinburgh. At the time of writing, she is still alive and living in Edinburgh. Her biographer, Robert Chalmers, commented: "It is arguable that had she been a man...she

Christina Miller (1899–) at her graduation. (Courtesy of Robert Chalmers.)

might have become one of the UK's first professors of analytical chemistry. At a time when analytical chemistry was practically nonexistent in UK universities she provided courses that would stand comparison with the best available today.... Her work was highly esteemed by many internationally renowned analysts."[15]

Emma Perry Carr (1880–1972)

Arguably the most famous woman chemical educator in the early part of this century was Emma Perry Carr. However, to discuss her life and work, we must place it within the context of Mount Holyoke College, where she spent most of her life. Mount Holyoke College, a women's liberal arts college in the small town of South Hadley, Massachusetts, has played a disproportionately large role for women chemists in the United States.[16] The original chemical impetus had come from Mary Lyon,[17] who founded Mount Holyoke Seminary in 1837 for the training of women teachers (it became a college in 1887). Lyon taught chemistry, both theory and laboratory work, as part of her belief that science gave crucial insights into the works of God. Her successor, Lydia Shattuck,[18] continued the traditional emphasis on chemistry, corresponding with the major scientific figures of the time and becoming active in scientific societies. She was one of a few women chemists to attend the 1874 Priestley Centennial meeting that led to the founding of the American Chemical Society, but she and the other women chemists were excluded when the historic official photograph was taken of the event.[19] Nellie Goldthwaite was the next motivating force of the chemistry department,[20] but the fourth incumbent, Carr, was to have the most influence.

The historians Carole Shmurak and Bonnie Handler have commented on the strength of chemistry at the College, as a result of Carr's efforts:

> When Lucy Goodale, Mount Holyoke Seminary class of 1841, wrote home to her parents, she called the seminary her "castle of science." She could not have known how apt that phrase would become in the next hundred years;...Nonetheless, Lucy Goodale's "castle" would become a citadel for women in science; a college that would produce more women who went on for doctorates in the physical sciences from 1910 to 1969, more women who obtained doctorates in chemistry from 1920 to 1980, and more women listed in the 1938 *American Men of Science*

than any other institution. Additionally, during the years 1927–1941, it would be twice as productive in publishing chemical research as any other liberal arts college (coeducational or single-sex) in the country.[21]

Emma Perry Carr was born in Holmesville, Ohio on 23 July 1880 to Edmund Cone Carr and Anna Mary Jack, a family of devout Methodists.[22] She was the third of five children, and the second eldest daughter. She came from an educated family, both her father and grandfather being country doctors, and at high school she was known as "Emmy the smart one". Following graduation, in 1898, Carr attended Ohio State University where she studied chemistry with William McPherson. Then in 1899, she moved east to enroll at Mount Holyoke College. Her first two years at Mount Holyoke were spent con-

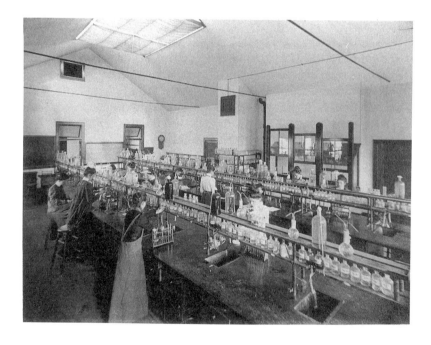

Mount Holyoke Chemistry Laboratory, ca. 1919. (Courtesy of Mount Holyoke College Archives and Special Collections.)

tinuing her studies of chemistry, but rather than graduate, she worked for the following three years as an assistant in the chemistry department. Granted a leave of absence from her work in 1904, she traveled to the University of Chicago where she completed a degree in chemistry in 1905. She then returned to Mount Holyoke where she was appointed instructor of chemistry.

Three years later, Carr again received a leave of absence, this time to obtain a doctorate at the University of Chicago. Her research topic, the study of aliphatic imido esters, was undertaken with Julius Stieglitz as supervisor. Stieglitz and Carr became lifelong friends. In 1910, she returned once more to Mount Holyoke, where she was promoted to associate professor, and then in 1913 to full professor and head of department.

Carr was a charismatic teacher, and with her colleagues Dorothy Hahn and Louisa Stephenson, she established a chemistry curriculum as rigorous as that at Yale University. She argued that the introductory chemistry course required the best and most experienced instructors if the enthusiasm for the subject was to be inculcated in the next generation. She had acquired this belief from one of her own mentors, Alexander Smith, as she herself commented:

> He emphasized always the paramount importance of laboratory work in the teaching of beginning chemistry and the necessity for having the most experienced teachers conducting the laboratory and weekly class discussions. It was rarely that Dr. Smith did not make the rounds of each laboratory section, asking questions, correcting misconceptions, and giving the student a sense of genuine interest and concern for his progress.[23]

Until 1935 Carr lived in a college dormitory, and this location enabled her to interact with her students at the dining table. She kept in touch with many of them for years after they graduated, and she was immensely proud of the 43 women who continued on to Ph.D. programs in chemistry. However, as Shmurak has noted, those bright chemistry students who married elicited a different response: "Although privately she felt each of these marriages a betrayal, she nonetheless accepted them graciously...."[24]

But her teaching abilities were only part of her efforts, for she believed strongly in the importance of research, even at a small teaching college. From reading "British and continental" chemistry journals, she noted that there was an interest in the relationship between ultraviolet absorption spectra and the structure of organic molecules.

However, there seemed to be no research in progress in North America. In 1913, she began work on the project with Dorothy Hahn, the College organic chemist, synthesizing the hydrocarbons. The first publication, which appeared in 1918,[25] was groundbreaking for the College in two ways: it established Mount Holyoke as a research institution, and it entrenched research work as part of the training of chemistry students. In fact, as Shmurak and Handler note, Carr was very keen on collaboration: "Faculty, graduate students, and honors undergraduate students all took part in this work, making each one feel 'a part of a very dedicated and hardworking group that was having fun together.'"[26]

Keen to learn the latest spectroscopic techniques, Carr wrote in 1919 to A. W. Stewart at the University of Glasgow. He suggested that she spend a year studying ultraviolet spectroscopy at the Queen's University in Belfast, Northern Ireland, and added a note on the friendliness of the department: "The Chemistry Department there has been run very much on the lines of a social club—I am not referring to their work, of course!—and I think they would make you feel at home at once. I merely mention these points because I have always believed that people do better work if they are comfortable...."[27] Her stay in Belfast further stimulated her interest in spectroscopy, but she did not anticipate where it would lead. In 1925, having acquired an expertise in spectroscopy, she was asked to become a Cooperating Expert on the preparation of the *International Critical Tables* (ICT). The ICT were to be the definitive collection of key chemical information, including spectroscopic data. Compiling the ICT involved searching through and evaluating spectral data from around the world. The work had to be coordinated with her two co-compilers, Victor Henri of the University of Zurich and Jean Becquerel of the Collège de France in Paris. For this task, Carr decided that she needed to be nearer to her collaborators, so she took a one-year leave of absence to travel to Zurich to work in Henri's laboratory and complete her contribution.

Following her return to the United States, she was asked by Morris S. Kharasch to collaborate on an investigation on electronic isomers or "electromers". In the 1920s, some researchers considered that organic isomerism could be a result of different electronic states rather than *cis/trans* geometrical isomerism. Carr and her new colleague Mary Sherrill, another graduate of Stieglitz in Chicago,[28] initially embraced the concept. As their studies continued, however, it became apparent from spectra of the high-frequency Schumann region of the ultraviolet that the electromer concept was in error.[29]

Emma Perry Carr (1880–1972). (Courtesy of Mount Holyoke College Archives and Special Collections.)

Following this research, Carr became interested in links between spectral absorption and heats of combustion of hydrocarbons. This work was not widely accepted, and its poor reception caused Carr some bitterness.

In 1935, Francis Garvan of the Chemical Foundation remarked on the absence of women chemists, and he offered a medal to recognize an American woman who had made a distinguished contribution to chemistry. Carr was opposed to any separate gender recognition, but she reluctantly agreed to be a member of the committee to choose that individual. To her embarrassment, during her absence, the remainder of the committee met and chose her as the first recipient of the award.[30]

With the outbreak of the Second World War, the women chemistry graduates of Mount Holyoke were again in great demand by industry, though once more the expectation was given that their employment would cease when the war was over. During this time, Carr, her chemistry colleagues, and the graduate and undergraduate chemistry students were pressed into service to attempt synthesis of quinine substitutes for the treatment of malaria. This was crucial work for, with the fall of the Far East to Japan, 90% of quinine supplies had fallen into Japanese hands. Though many analogs were synthesized at Mount Holyoke, none were to prove effective.

Carr retired from teaching in 1946, though she kept active as a popular speaker at universities and clubs into her 70s. From 1935 to 1961, Carr shared a house on campus with Sherrill, and following Sherrill's retirement in 1954, the two traveled widely together. Carr died on 7 January 1972. At a memorial service, a mourner remarked that "it was a resistant person who could fail to share her enthusiasm, whether for science, for politics, for her family, for pi electrons, for baseball, or for the circus."[31]

Mary Fieser (1909–)

Textbook writing has always been a popular role for women chemists[32] as we saw with Jane Marcet in Chapter 3, and one of the most successful writers in modern times has been the organic chemist, Mary Fieser.

Born Mary Peters in Atchison, Kansas, Fieser came from a well-educated family.[33] Her paternal grandfather had been president of Midland College, a small Lutheran college, and her maternal grandmother had graduated from a ladies' seminary. Her mother, Julia Clutz, had taken graduate studies at Goucher College, and her father,

Robert Peters, was a professor at Midland College (and later, at Carnegie-Mellon University). Fieser was one of two children, her sister Ruth becoming a professor of mathematics after obtaining a Ph.D. at Harvard University.

Fieser was strongly influenced by her maternal grandmother, who impressed upon her that girls, particularly, should be busy every minute. Fieser remarked: "She sat us down at eight in the morning. If we stopped during the day to catch our breath, she would remind us that 'the devil finds work for idle hands.'"[34] As well, she was taught that little sleep was really necessary, so Fieser adopted the habit of going to sleep at 10 P.M. and rising at 2 A.M. Her father was quite traditional, wanting his daughters to "dress in frills" and "look sweet."[35] However, he did permit her to indulge in card games, such as whist. Excelling at them, Fieser found that games were an outlet for her intensely competitive nature.

The family moved to Harrisburg, Pennsylvania, where Fieser attended a private girls' school. From there, she was accepted into Bryn Mawr College, graduating with a B.A. degree in chemistry in 1930. Her chemistry instructor throughout her time at Bryn Mawr was Louis Fieser, and like other students, she found his chemistry lectures inspiring, particularly his emphasis on experimental rather than theoretical aspects. When Louis Fieser moved to Harvard in 1930, Mary Fieser enrolled as a graduate student there. She spent half of her time doing research with Louis Fieser[36] and the other half taking courses. Harvard was not very welcoming to women at that time; in particular, her professor for analytical chemistry, Gregory Baxter, would not allow her to attend the regular laboratories. Instead, she was told to conduct her experiments, unsupervised, in the deserted basement of an adjoining building with only Baxter's woman research student, Evelyn Emma Behrens, for company. In 1932, she completed the requirements for an A.M. in organic chemistry.

Fieser decided on an A.M. degree rather than a Ph.D. as the result of a proposal of marriage from Louis Fieser, which she accepted the same year as the completion of her master's. With the antifemale attitude at Harvard, Fieser felt more secure continuing research as a spouse than by obtaining a doctorate. She commented, "I could see I was not going to get along well on my own, [but after marriage] I could do as much chemistry as I wanted."[37] Following marriage, Fieser became an unpaid member of her husband's research staff. Later, she was formally appointed as a research associate (without pay).

During the 1930s and 1940s, Mary Fieser was Louis Fieser's "prime co-worker", studying the chemistry of quinones and of natural products, particularly steroids. During the war, the Fiesers, like the students at Mount Holyoke, were involved in the search for antimalarial substitutes for quinine. They found a promising candidate, lapinone, but unfortunately it could only be used by injection.[38]

It was in 1942 that the Fiesers started writing texts. Mary Fieser suggested this as she was concerned that, with Louis Fieser's heavy research commitments at the time, he might lose touch with teaching. After doing background reading, she offered her notes to her husband, who then wrote the text. However, she collected material at a faster rate than he could process it, and at his suggestion she began to compose some of the chapters herself. This book, *Organic Chemistry*,[39] appeared in its first edition in 1946. Following its success, they wrote *Introduction to Organic Chemistry*,[40] *Basic Organic Chemistry*,[41] *Style Guide for Chemists*,[42] *Advanced Organic Chemistry*,[43] *Topics in Organic Chemistry*,[44] and the classic work, *Reagents for Organic Synthesis*.[45] This last work has been a continuing series that she wrote alone (or for some volumes, with other collaborators) following the death of Louis Fieser in 1977. The writing involved 15-hour days, scanning the organic chemistry literature, looking for original reactions that could be used in organic synthesis. Thus, like many of her predecessors from Bryn Mawr, Fieser has devoted her life full-time to science. At the time of writing, Fieser is living in Massachusetts.

Hélène Metzger (1889–1944)

We are ending this chapter with the life of one of the great historians of chemistry. Hélène Emilie Metzger was born on 26 August 1889, near Paris, the daughter of Paul Bruhl and Eugénie Emilie Adler.[46] Sadly, when Metzger was two years old, her mother died while giving birth to a second daughter, Louise. Six years later, Paul Bruhl remarried, but Metzger never accepted her step-mother, a factor which accounted for her very introverted nature and independence of spirit. In a letter to the science historian, George Sarton, she commented that her father "had refused, following the ideas of his time, to allow his socialist (or almost) daughters to pursue independent careers."[47] Nevertheless, despite parental disapproval, Metzger proceeded to obtain a higher-level high-school graduation diploma (the *Brevet Supérieur*) and applied to attend the Sorbonne.

She chose to study the structures of crystals with professor Frédéric Wallerant, her interest probably deriving from the fact that the Bruhl family business was in diamonds, pearls, and precious stones.

In 1912, she received a diploma for advanced studies on the basis of her crystallographic analysis of lithium chlorate. The following year, she married Paul Metzger, a professor of history and geography at Lyon, but sadly, he was killed in one of the first battles of the First World War. A widow at 25, Hélène Metzger devoted the rest of her life to research.

With a wide-ranging curiosity and an instinct for philosophy, she turned her interest toward the history of science. Her abilities were apparent from her doctoral thesis on the origins of crystallography, submitted in 1918. In her thesis, she showed how the study of crystals slowly became differentiated from mineralogy, physics, and chemistry, until at the beginning of the eighteenth century, crystallography became an independent science. This focused her attention on the history of chemistry, particularly the concepts of chemistry in France from the beginning of the seventeenth century through to the end of the eighteenth century. But she did not want to just collate historical material; she wanted to examine the underlying philosophy of science and develop her own perspective on the advances in chemistry. In particular, Metzger considered that the understanding of scientific progress required the examination of "false" theories as well as those that were "true." For example, in Metzger's view, the alchemical research of Nicholas Lémery had played a major, and until then overlooked, role in the development of modern chemistry. The first volume of her work appeared in 1923, and it won her the Binoux Prize of the French Academy of Sciences.[48]

Following this success, she wrote a popular history of chemistry.[49] The science historian Bernadette Bensaude-Vincent describes how the book provided a different perspective on the subject: "Although she is writing for the general public, she [Metzger] refuses to stoop to writing the usual 'history of great men' or to painting a series of portraits. ... The historian is concerned with ideas, with the intellectual system they form at any given moment, and with how they develop. Individuals play only a minor role by producing variants...."[50]

During the 1930s, Metzger wrote numerous books and articles on aspects of the history of early modern chemistry, each one of which was a penetrating analysis.[51] She was active in history of science organizations, being appointed administrator-treasurer of the Académie Internationale d'Histoire des Sciences in 1929. In 1939, she was placed in charge of the history of science library of the Centre International de Synthèse, and she became the secretary of the Groupe Français d'Historiens des Sciences. Suzanne Delorme has commented that Metzger enlivened discussions "with her subtle and often ironic remarks, which were always pertinent and erudite, if somewhat disconcerting in their impulsiveness."[52]

In 1941, Metzger moved from German-controlled Paris to Vichy-controlled Lyon. She could have kept a low profile in Lyon, not drawing attention to her Jewish background. But this was not Metzger's way. Instead, she was proud of her ancestry, and she started work at the Bureau d'Études Israëlites on a study of Jewish monotheism. It was three years later, close to the end of the war, on 8 February 1944, that she was arrested on the basis of her Jewish background during a police raid on her boarding house. Initially held in the Montluc fort, she was transferred to the town of Drancy on 20 February. She was included in a group of 1501 prisoners who left Drancy in convoy 69 on 7 March, arriving at Auschwitz on 10 March 1944. All were taken directly to the gas chambers except for a few young and strong individuals.[53] Thus the life of one of the great historians of chemistry came to a tragic end.

Chapter 10

Epilogue

*The Dawn and Dusk of an Era—The Situation in Chemistry—
Causes for the Decline—The Second World War and Its After-
math—Women of Color in the Chemical Sciences—The Support
Roles of Women in Chemistry—Synthesis—Concluding Remarks*

The Dawn and Dusk of an Era

At the end of the nineteenth and the beginning of the twentieth centuries, the growth of women's participation in higher education was
phenomenal. In the context of the United States, Patricia Graham
has commented:

> By 1880 women constituted 32 percent of the student
> body; by 1910 almost 40 percent. A decade later, 1919–
> 1920, women were 47 percent of the undergraduate
> enrollment, nearly as much as their proportion in the
> population. The decade of the twenties was critical for
> educated women. During that ten year period, women
> achieved their highest proportion of the undergraduate
> population, of doctoral recipients, and of faculty mem-
> bers. A record 32.5 percent of college presidents, profes-
> sors, and instructors were women when the decade closed
> in 1930.[1]

The 1920s were to be a zenith for academic women,[2] a level of attain-
ment that was not to be reached again until well into the second half
of the twentieth century.

For the students of the time, particularly those entering the work-
force during the First World War, everything seemed possible. Only

later did the realities of life become apparent. Marjorie Nicolson, a 1914 graduate of the University of Michigan, described her feelings:

> We of the pre-[First World] war generation used to pride ourselves sentimentally on being the "lost generation," used to think that, because war cut across the stable path on which our feet were set, we were an unfortunate generation. But as I look back upon the records, I find myself wondering whether our generation was not the only generation of women which ever found itself. We came late enough to escape the self-consciousness and belligerence of the pioneers, to take education for granted. We came early enough to take equally for granted professional positions in which we could make full use of our training. This was our double glory. Positions were everywhere open to us; it never occurred to us at the time that we were taken only because men were not available.... The millenium had come; it did not occur to us that life could be different. Within a decade shades of the prison house began to close, not upon the growing boy, but upon the emancipated girls.[3]

Specifically in science, women's experiences were much the same. In 1898, Henrietta Bolton had described the optimistic future for women pursuing science careers.[4] She contrasted the "clinging, fainting, willowy heroine, dear to the hearts of our grandmothers" with the "alert, athletic, breezy woman" of the end of the nineteenth century who was ready to make great discoveries in science. Yet here too, the 1920s were to be a turning point, as Betty Vetter has described: "Once before, women made a dent in the sturdy armor of the male science community. During the decade of the 1920's, women earned 12 of every 100 Ph.D.'s awarded in science and engineering, but this was a higher proportion than they ever would again until 1975."[5]

The Situation in Chemistry

The enrollment of women in chemistry programs followed the same pattern as science in general: a rapid increase toward the start of the twentieth century, peaking for undergraduate degrees in the 1910s and 1920s.[6] For each college, the year of maximum enrollment differed slightly. For example, at the women's college of Bryn Mawr, the number of chemistry majors reached maxima over the years 1914 to 1916 and again from 1919 to 1922 where the maximum chemistry enrollment at Mount Holyoke College was for the year 1922. Similar enrollment patterns applied to coeducational universities, such as the University of Wisconsin-Madison, where women made up 22% of chemistry classes in 1920, dropping to 10% in 1925. This peaking of enrollment of women in chemistry (and the other sciences) during the 1920s was not only true in the United States but also in Britain and other Western countries.

Kenneth Everett and Will DeLoach[7] showed a parallel trend for U.S. doctorates conferred on women. They observed a drop in U.S. women chemistry doctorates after 1932 (one would, of course, expect the doctorate maximum to occur some years after the B.S. zenith). This decrease was in actual numbers, but it is particularly striking in terms of percentage of doctorates awarded. A maximum was reached in 1929 when 10% of chemistry Ph.D. degrees were awarded to women, dropping to 5% by 1933, and reaching a minimum of 2% in the 1940s. The 10% figure was reached again in the abnormal year of 1946 (which came after five years in the 2–4% range) and not regained until 1972, when women started to consistently exceed the 10% figure.

Causes for the Decline

There are several factors that can be cited for this decline. First of all, the interwar years saw an increase in discrimination against women. Before the First World War, women were tolerated in male professions insofar as they were not identified as a threat but more as a curiosity. During the First World War, however, women filled a significant portion of skilled occupations. As a result, with the end of the war, as Elizabeth Roberts commented: "They [women] were often regarded with open hostility by men who had realised for the first time that women were fully capable of carrying out jobs previously perceived as men's, thus presenting a real challenge."[8]

The Canadian historian Veronica Strong-Boag noted the general rise in prejudice in Canada against academic women during the 1930s, with some of the new university scholarships being reserved exclusively for men.[9] The hostility was so widespread at the University of Alberta, Canada, that the head of the Woman-Hater's Club was elected president of the student union in 1935. The antifeminist attitude was also prevalent in Italian universities. One popular student song contained the line: "We don't want women at school, we want them naked laid out on a couch."[10] The misogynistic trend was

related by Strong-Boag to the change in the image of women from
the independent "flapper" of the 1920s to the dependent "moppet"
of the 1930s. The feminist author Susan Faludi also remarked on this
change of attitudes toward women between the 1920s and 1930s:
"Postfeminist sentiments first surfaced, not in the 1980s media, but in
the 1920s press. Under this barrage, membership in feminist organizations soon plummeted, and the remaining women's groups hastened to denounce the Equal Rights Amendment or simply converted
themselves to social clubs."[11]

But the problem was not just a change in societal attitudes.[12] In
the early period of access to university, women were taking degrees to
satisfy a zest for knowledge: a degree was a goal in itself.[13] As the twentieth century progressed, the question of careers became more
important and it became apparent that there were few acceptable
employment possibilities with a science degree, whereas an arts
degree opened the doors to many women-accessible careers such as
teacher, social worker, librarian, clerical worker, and saleswoman.
Those women who graduated with degrees in chemistry usually found
themselves channeled into "women's fields", such as home economics, or else given "dead-end" support positions.[14]

It was not only individuals but also some governments that were
trying to restrict the advancement of women in the high-unemploy-
ment interwar period. The education of women was regarded as leading to the displacement of skilled male workers (hence a higher
reported unemployment rate) and to a reduction in birthrate, both
implications being considered as social evils. As a result, many Western countries passed laws banning the employment of married
women in professional positions. The greatest efforts to combat this
perceived problem were made in Germany, where the "threat" of
young women choosing intellectual rather than family-rearing activities was taken very seriously by the Nazi government. A law passed in
1934 required female (but not male) high-school graduates to first
perform a six-month "labour service" before they could apply to a university. This was intended primarily as an obstacle that would curb, or
at least delay, women's enrollments at the institutions of higher learning. As Jacques Pauwels remarked: "Moreover, the lives of the 'labor
maids' (*Arbeitsmaiden*) in the camps was far from pleasant.... The pros-
pect of six months in a Spartan labor camp far from home, of hard
work in the fields, inexorable discipline, and minimal comfort was
undoubtably a powerful deterrent to many young women who would
normally have embarked on an academic career."[15] Obviously,
women did not disappear from chemistry in the 1930s and 1940s, but,
in general, they did become more marginalized.

In the physical sciences, another reason for a drop in women's participation was a change in the nature of research activity. Earlier in this book, we argued that women tended to enter the fields of chemistry that had collaborative, supportive atmospheres. However, the goal of the Nobel Prize and other forms of recognition and fame seemed to change the nature of the research endeavor to a fiercely competitive environment,[16] such an atmosphere being very discouraging to women.[17] For example, when Joan Freeman arrived at the Cavendish Laboratory, Cambridge, in 1946, she was told: "Progress would be painful and intellectually demanding.... Worthwhile results would not come easily. Far too many people were flocking into the field.... It would be a case of survival of the fittest."[18] What a change this was from the pleasant convivial atmosphere that had been described 40 years earlier, and surely an additional factor contributing to the decline of women's first major participation in science.

The Second World War and Its Aftermath

The onset of war once more opened up opportunities for women, but they were to be temporary openings, as had been the case for the preceding generation of women chemists. When the men began to be drafted from college campuses, women were hired as temporary faculty, without possibility of tenure. Thus between 1942 and 1946, the number of women chemistry faculty in the United States increased from 485 to 1585.[19]

In the chemical industry, the expected demand for women chemists was far less than anticipated as employers chose women applicants only as a last resort. The science historian Margaret Rossiter has quoted a speech by Walter J. Murphy, editor of *Industrial and Engineering Chemistry*, with her own additional comments in parentheses:

> They [the women audience] should realise that women, in general, had so many weaknesses that an employer would find it hard to take them seriously. Almost any male chemist was preferable to any woman, since the women were known to be unable to do teamwork as well as (all) men; unwillingness to work under the supervision of other women; lacking in aggressiveness (which seemed to mean originality, perseverance, dedication, and eagerness to get ahead) and in the focused type of imagination that (all) men have![20]

Alice Keubel mentioned other problems of hiring women chemists. According to her survey of employers, they were reluctant to hire women because of the need to provide women's washrooms and because "men covet their privilege to swear, and on hot summer days they like to work in their undershirts—privileges supposedly impossible when a woman enters the lab."[21] Lois Woodford raised the issue of "matrimonial mortality" (resignation following marriage), which made women less valuable employees than men.[22] The one, somewhat dubious, advantage of hiring a woman was her willingness to work for as little as one-half of the salary of a male chemist. In spite of these problems, women chemists were urged to persevere and demonstrate their ability to excel in their work.[23]

With the end of the war, the returning men expected to reoccupy their former positions. Thus one woman chemist advised her colleagues to accept "cheerfully" any demotion resulting from the return of the former (male) occupant of their position.[24] But the loss of "temporary" jobs was not the only problem. Two women chemists at the Hercules Powder Company of Delaware complained that women were not promoted at chemical companies, whatever they had accomplished. They argued that nothing would change until the management considered "the woman chemist as a chemist rather than as a woman."[25] However, many women chemists, such as Emma Perry Carr (whose biography is given in Chapter 9), believed that such protests were more likely to lead to reprisals rather than an improvement in the situation. Instead, she considered that the best response was for professional women chemists to perform "a rather higher quality of work than might be expected of a man whose choice of profession is assured."[26]

Another woman chemist of the time, Cornelia Snell, argued that women chemists should seek niches in which they could excel, such as microanalysis where "[w]omen have a delicacy of manipulation in handling small objects."[27] However, she felt the best advice for the woman chemist was to avoid competition with men chemists by choosing a career such as technical secretary, exhorting her readers that the potential for advancement would be much greater than that of a woman laboratory chemist.

By the end of the first half of the twentieth century, the situation for women chemists was far from hopeful. Rossiter has summarized the plight of women scientists in general: "They had lost ground in employment and advancement in academia, government and industry, and yet, except for a few individuals, they were too afraid even to protest."[28] It was to be the second half of the twentieth century before women chemists were to regain the lost ground and, indeed, to be widely recognized as chemists as well as women.

Women of Color in the Chemical Sciences

We mentioned in Chapter 1 that there had been Chinese women alchemists throughout the historical record. However, chemistry, until recently, has been not only a male science, but also a "white" science. Other sciences opened up to women of color earlier and/or more dramatically than chemistry. In biology, for example, Afro-American Roger Arliner Young (1899–1964)[29] broke the ground for her successors, becoming a researcher at the prestigious Marine Biological Laboratory at Woods Hole, Massachusetts, in 1927, prior to obtaining a Ph.D. from the University of Pennsylvania in 1940. Among women nuclear physicists, one of the brightest stars has been Chinese-born Chien-Shiung Wu (1912–1997).[30] Wu completed her doctoral studies at the University of California at Berkeley and later became a full professor at Columbia University.

The earliest recorded Afro-American woman chemist was Mary Elliott Hill (1907–1969), who obtained a B.S. from Virginia State College in 1929 and an M.S. from the University of Pennsylvania in 1941.[31] It was not until 1948 that Marie Maynard Daly was to become the first Afro-American woman to receive a doctorate in a chemical field; she was awarded her Ph.D. in biochemistry from Columbia University.[32] However, few followed in their pioneering footsteps until the latter part of the twentieth century. In many of the biographical accounts in this compilation, a mentor was a vital stimulus, and perhaps it was a lack of such a mentor in chemistry during the time frame of our study that was the problem. This was unlike the situation in mathematics, where Lee Lorch at Fisk University encouraged many Afro-American women to pursue mathematical careers at a time when it was unpopular to champion the cause of either women or Afro-Americans.[33] Of Lorch's protégées, many proceeded to mentor the next generation of Afro-American women mathematicians.

The Support Roles of Women in Chemistry

To show that women have made significant contributions to chemistry, we have to a certain extent used the "great women" theme, though we have endeavored to weave in appropriate contextual material. But this approach does not truly reflect the important roles of the subordinates as Ruth Hubbard has remarked:

> The laboratory workers—the people who work with their hands—are the ones who perform the operations and

make the observations that permit hypotheses and ideas to become facts. Often they are the ones who produce the substrata of observations out of which the new ideas emerge, that the laboratory chief then puts out as his, or occasionally her own.[34]

And it is in the support role that, historically most women were found. As Hubbard continued:

Women have played a very large role in the production of science—as wives, sisters, secretaries, technicians, and students of "great men"—though usually not as named scientists.[35]

Though some women were probably content with their subordinate status, we must realize that for other women over the centuries, their dream to become a chemist was dashed by societal opposition or financial limitations and, instead, they had to revert to the conventional life expected for a woman of their time.

Synthesis

Before the conclusion of this book, it is appropriate to review some of the common threads that run through the lives of the women chemists we have discussed.

A factor not mentioned until now is the incredible longevity of many of these pioneers. In the French salon era, the average lifespan of the three women chemist-assistants was close to 80 years (excluding du Châtelet who died in childbirth), longevities far exceeding the norm of the time. The field with the least average lifespan was that of radioactivity, and, as we discussed, the early demise of some of these women researchers was a result of the casual handling of radioactive materials. Even then, there were exceptions such as Lise Meitner (1878–1968) and Ellen Gleditsch (1879–1968), two women who lived over almost identical long lifespans. The record longevity to date is the biochemist Helen Dyer, who has now continued into her second century.[36] To what do we attribute these long lifespans? In our view, it is the enthusiasm and dedication of these women that gave them a continuing reason (or even obsession) to live. Many of the women worked at their chosen field almost until the day that they died. Thus Dyer continued to review grant proposals and technical papers until the age of 95, and the industrial chemist Florence Wall (1893–1988), became honorary activist of the American Institute of Chemists at the age of 93. An example from the field of radioactivity was the "polonium woman," Elizabeth Róna (1890–1981).[37] Róna retired from her position at the Oak Ridge Associated Universities at the age of 75 and then accepted an offer from the University of Miami where she started a new research interest, that of geochronology. Retiring again at age 82, she wrote her autobiography and traveled widely during her remaining years.

The overriding dedication of these women comes through in their deeds and words time and time again. Icie Macy Hoobler provided one such example in her willingness to suffer nephritis rather than relinquish her laboratory in a building that lacked women's restroom facilities. From an earlier epoch, Émilie du Châtelet was determined to work night and day on her manuscript for fear that it would not be completed before her death. And then there were the Russian women, such as Yulya Lermontova, whose enthusiasm for learning led them to travel far from their native country in search of a university that would accept them. In some cases, this quest necessitated finding a temporary marriage partner so that the woman could leave Russia in the first place. Vera Bogdanovskaia probably provided the most unusual way of expressing dedication to chemistry by insisting upon her own chemistry laboratory as a condition of marriage. On a larger scale, one admires the devotion of the "canary girls" in the First World War, who risked their health and, in some cases, gave their lives, for the war effort. The greatest dedication of all must be Rosalind Franklin's wish to work on the structure of the live polio virus once she knew that she was terminally ill.

Few of the women found their scientific progress obstacle-free. Until the late nineteenth century, women were deprived of access to higher education, making it extremely difficult for them to make any significant contribution to chemistry. When that hurdle was removed, women planning careers in chemistry faced the opposition of prominent scientists of their time. Apart from the unique case of Marie Curie, even the most talented women made little progress up the institutional hierarchy. Nobel Laureate Dorothy Hodgkin spent much of her life in the lowest possible rank at Oxford University, and Gerti Cori and Maria Goeppert-Mayer were treated essentially like academic wives until they each received Nobel Prizes. Moreover, these individuals were the fortunate ones who did receive significant recognition for their work during their lifetime, unlike others such as Lise Meitner and Rosalind Franklin.

Perhaps the most pervasive feature of our biographical accounts has been the importance of mentors. We have seen the role of men-

tors for the chemist-assistants in the French salon period and again in
Lord Rayleigh's encouragement of Agnes Pockels. During the early
twentieth century, we showed the valuable mentoring roles of the
crystallographers W. H. Bragg, W. L. Bragg, and J. D. Bernal; the biochemists F. G. Hopkins and L. Mendel; and the atomic scientists E.
Rutherford and S. Meyer. What is significant about these mentors is
not only that they supported the rights of women at a time when it
was not fashionable to do so, but that they showed exceptional kindness to their students. The importance of mentoring for women
chemists raises the rather depressing question of how many women
had the enthusiasm and desire to pursue a career in chemistry, yet
failed to find a mentor and "fell by the wayside."

The question of marital status is also important. Many of the
women, such as Agnes Pockels, Ellen Gleditsch, Lise Meitner, and
Christina Miller, remained single. Their work was their life. One of
Marie Curie's former students, the unmarried Swedish chemist Eva
Ramstedt (1879–1974), wrote to Curie: "As for me, I have been
unfaithful to science, never finding time for research."[38] In fact, it was
argued at the time that women who excelled academically should
regard it an honor and a duty to devote their lives to knowledge as if it
were a religious vocation and to avoid even the thought of marriage:
"Civilization rests upon dedicated [women's] lives, lives which
acknowledge obligation not to themselves or to other single persons,
but to the community, to science, to art, to the cause."[39] Yet there was
no similar expectation of male celibacy and the large majority of the
male figures in our accounts, such as the Braggs and Rutherford,
were married. But of course, there was no equivalency. For a woman,
a husband represented commitments additional to research whereas
for a man, a wife provided a means of lightening commitments outside of research. To illustrate this point, the sociologist Terri Apter,
mentions a female academic who was asked why she remained single.
The professor replied that she would have certainly married, had she
found a suitable wife.[40]

For those who did marry, a significant proportion did not have
children. In the cases of Marie Lavoisier and Claudine Picardet, the
lack of children made their laboratory presence acceptable in French
scientific society. Also, the avoidance of childbirth probably contributed to their longevity, particularly when we consider the fate of du
Châtelet. In almost every case, women married co-researchers, other
examples being Marie Curie, Irène Joliot-Curie, Ida Noddack, and
Mary Fieser. Perhaps the most fascinating cases were those of Kathleen Lonsdale and Dorothy Hodgkin, who actually married academics from other disciplines. Their respective spouses recognized the

brilliance of their wives, and the men became virtual "househusbands" in their willingness to shoulder child-raising burdens. Finally, we have the category of women for whom marriage spelled the end of their research careers, such as Harriet Brooks and Edith Willcock.

Whatever the common factors, the life of each one of the women represents a unique path, one that for its time must have been daunting and, at times, overwhelming. It is unfortunate that we have so little record of the feelings of these individuals. In fact, it was only Marie Meudrac and Elizabeth Fulhame who bravely expressed their feminist sentiments in print. If only, we could have known more of what these women chemists thought, as well as what they did.

Concluding Remarks

Our historical coverage has taken us from the beginning of recorded human history to the end of the first period of significant participation of women in chemistry. It is an appropriate place to terminate this study, for the 1950s marked the end of an era. Access to higher education and the possibility of a career in chemistry again became acceptable for women. Even the most recalcitrant universities opened their doors to women, and laws were enacted in most Western countries to prevent overt discrimination. The "second wave" of women in science arrived and flourished. Will this wave prove to be more permanent? If women chemists are to play a full role in determining the nature and culture of chemistry in the twenty-first century, their proportions have to exceed a "critical mass",[41] and for many institutions, such a situation is still in the future.

References and Notes

Chapter 1

1. Chemical technology, such as metal smelting and brick making, can be traced back to prehistory. *See* Keyser, P. T. "Alchemy in the Ancient World: From Science to Magic." *Illinois Classical Studies* **1990**, *15*, 353–378; Salzberg, H. W. *From Caveman to Chemist: Circumstances and Achievements;* American Chemical Society: Washington, DC, 1991.

2. Levey, M. "Some Chemical Apparatus of Ancient Mesopotamia." *J. Chem. Educ.* **1955**, *32*, 180–183.

3. Levey, M. "Perfumery in Ancient Babylonia." *J. Chem. Educ.* **1954**, *31*, 373–375.

4. Tyldesley, J. *Daughters of Isis: Women of Ancient Egypt;* Penguin: London, 1994.

5. Davis, T. L. "Primitive Science, the Background of Early Chemistry and Alchemy." *J. Chem. Educ.* **1934**, *11*, 3–10.

6. Pomeroy, S. B. *Goddesses, Whores, Wives, and Slaves: Women in Classical Antiquity;* Schocken Books: New York, 1975.

7. Carlson, A. C. "Aspasia of Miletus: How One Woman Disappeared from the History of Rhetoric." *Women's Stud. in Commun.* **1994**, *17(1)*, 26–44.

8. Alic, M. *Hypatia's Heritage: A History of Women in Science from Antiquity Through the Nineteenth Century;* Beacon Press: Boston, MA, 1986.

9. Alic, M. "Women and Technology in Ancient Alexandria: Maria and Hypatia." *Women's Stud. Int. Quart.* **1981**, *4*, 305–312.

10. Patai, R. "Maria the Jewess: Founding Mother of Alchemy." *Ambix* **1982**, *29*, 177–197.

11. Cited in Patai, "Maria the Jewess," p 178.

12. *See*, for example, Holmyard, E. J. "An Alchemical Tract Ascribed to Mary the Copt." *Archeion* **1927**, *8*, 161–168.

13. Meislich, E. K. "The Eve of Chemistry." *CHEMTECH* **1978**, *8*, 588–592.

14. Dzielska, M. *Hypatia of Alexandria;* Harvard University Press: Cambridge, MA, 1995; Rist, J. "Hypatia." *Phoenix* **1965**, *19(3)*, 214–225; Deakin, M. B. "Hypatia and Her Mathematics." *Am. Math. Monthly* **1994**, *101*, 234–243.

15. Dubs, H. H. "The Beginnings of Alchemy." *Isis* **1947**, *38*, 62–86.

16. Meschel, S. V. "Teacher Keng's Heritage: A Survey of Chinese Women Scientists." *J. Chem. Educ.* **1992**, *69*, 723–730.

17. Davis, T. L. "Ko Hung (Pao P'u Tzu), Chinese Alchemist of the Fourth Century." *J. Chem. Educ.* **1934**, *11*, 517–520.

18. Nriagu, J. O. "A Precious Legacy." *Chem. Br.* **1994**, *30*, 650–651.

19. Davis, T. L.; Wu, L-C. "Ko Hung on the Yellow and the White." *J. Chem. Educ.* **1936**, *13*, 215–218.

20. Cited in Needham, J. *Science and Civilization in China: Vol. 5, Chemistry and Chemical Technology;* Cambridge University Press: Cambridge, UK, 1976; p 170.

21. Huff, T. E. *The Rise of Early Modern Science: Islam, China, and the West;* Cambridge University Press: Cambridge, UK, 1993.

22. Flanagan, S. *Hildegard of Bingen 1098–1179: A Visionary Life;* Routledge: London, 1989.

23. *Hroswitha of Gandersheim: Her Life, Times, and Works, and a Complete Biography;* Haight, A. L., Ed.; The Hroswitha Club: New York, 1965.

24. Noble, D. F. *A World Without Women: The Christian Clerical Culture of Western Science;* Oxford University Press: New York, 1992; p 136.

25. In 1293, Pope Boniface decreed that nuns be totally cloistered, hence adding to their isolation and inability to contribute to scientific progress. *See* Anderson, B. S.; Zinsser, J. P. *A History of Their Own: Women in Europe from Prehistory to the Present;* Harper & Row: New York, 1989; Vol. 1, p 193.

26. Luchaire, A. (Krehbiel, E. B., trans.) *Social France at the Time of Philip Augustus;* Harper & Row: New York, 1967; p 71.

27. Cited in Noble, *A World Without Women;* pp 154, 236. Heisch has noted "...the Queen, having established herself as an exceptional woman, did nothing to upset and interfere with male notions of how the world was or should be organized....[T]here is no indication that she encouraged ladies of the court to pursue learning after her example." Heisch, A. "Queen Elizabeth I and the Persistence of Patriarchy." *Feminist Rev.* **1980**, *4*, 45–55.

28. Wertheim, M. *Pythagoras' Trousers: God, Physics, and the Gender Wars;* Random House: New York, 1995. ©1995 by Random House, Inc. Reproduced by permission.

29. One of the few women to obtain recognition for her intellect in that period was the Quaker aristocrat, Anne Finch, Viscountess Conway (1631–1679). *See* Merchant, C. *The Death of Nature: Women, Ecology, and the Scientific Revolution;* Harper & Row: San Francisco, CA, 1980, pp 255–268; Fraser, A. *The Weaker Vessel;* Alfred A. Knopf: New York, 1984, pp 345–357.

30. Eastlea, B. *Witch Hunting, Magic and the New Philosophy: An Introduction to*

Debates of the Scientific Revolution 1450–1750; Harvester Press: Sussex, UK, 1981; Barstow, A. L. *A New History of the European Witch Hunts;* Harper: San Francisco, CA, 1994.

31. Mozans, H. J. (pen name of Zahm, J. A.). *Woman in Science;* reprint; University of Notre Dame Press: Notre Dame, IN, 1991; p 240. © 1991 by the University of Notre Dame Press. Used by permission of the publisher.

32. Herzenberg, C. L. *Women Scientists from Antiquity to the Present: An Index;* Locust Hill Press: West Cornwall, CT, 1986, p xxviii.

33. Cited in Gilchrist, C. *Alchemy: The Great Work;* Aquarian Press: Wellingborough, UK, 1984; p 88.

34. Cited in Alic, *Hypatia's Heritage;* p 59.

35. Phillips, P. *The Scientific Lady: A Social History of Women's Scientific Interests 1520–1918;* Weidenfeld & Nicolson: London, 1990; pp 52–56. *See also* Ilsley, M. H. *A Daughter of the Renaissance, Marie le Jars de Gourney, Her Life and Works;* Mouton: The Hague, Netherlands, 1963.

36. Bell, S. G. "Christine de Pizan (1364–1430): Humanism and the Problem of the Studious Woman." *Feminist Stud.* **1976,** *3(3/4),* 173–184; Willard, C. C. *Christine de Pizan: Her Life and Works;* Persea Books: New York, 1984; *The City of Scholars: New Approaches to Christine de Pizan;* Zimmerman, M.; De Rentiis, D., Eds.; de Gruyter: Berlin, Germany, 1994.

37. Phillips, *The Scientific Lady;* p. 56. *See also* Irwin, J. "Anna Maria Van Shurman: From Feminism to Pietism." *Church Hist.* **1977,** *46,* 48–62.

38. Bishop, L. O.; DeLoach, W. S. "Marie Meudrac—First Lady of Chemistry?" *J. Chem. Educ.* **1970,** *47,* 448–449.

39. Meudrac, M. *La Chymie charitable et facile, en faveur des dames;* se vend rue de Billettes: Paris, 1666.

40. From Meudrac, *La Chymie charitable et facile, en faveur des dames;* p xxxv, English translation; cited in Bishop, L. O.; DeLoach, W. S. "Marie Meudrac—First Lady of Chemistry?" *J. Chem. Educ.* **1970,** *47,* 448–449.

41. From Meudrac, *La Chymie charitable et facile, en faveur des dames;* p xxxii, English translation; cited in Bishop, L. O.; DeLoach, W. S. "Marie Meudrac—First Lady of Chemistry?"

42. Elena, A. "In Lode Della Filosofessa di Bologna: An Introduction to Laura Bassi." *Isis* **1991,** *82,* 510–518; Findlen, P. "Science as a Career in Enlightenment Italy: The Strategies of Laura Bassi." *Isis* **1993,** *84,* 441–463; Logan, G. B. "The Desire to Contribute: An 18th Century Italian Woman of Science." *Am. Hist. Rev.* **1994,** *99,* 785–812.

43. Truesdale, C. "Maria Gaetana Agnesi." *Arch. Hist. Exact Sci.* **1989,** *40,* 113–142.

44. Dobbs, B. J. T. "From the Secrecy of Alchemy to the Openness of Chemistry." In *Solomon's House Revisited: The Organization and Institutionalization of Science;* Frangsmyr, T., Ed.; Science History: Canton, MA, 1990; pp 75–94.

Chapter 2

1. Roberts, L. "Filling the Space of Possibilities: Eighteenth Century Chemistry's Transition from Art to Science." *Sci. Context* **1993,** *6,* 511–553.

2. Partington, J. R. *A History of Chemistry;* Macmillan: London, 1962; Vol. 3; Melhado, E. M. "Chemistry, Physics, and the Chemical Revolution." *Isis* **1985,** *76,* 195–211; McEvoy, J. G. "The Chemical Revolution in Context." *Eighteenth Century: Theory and Interpretation* **1992,** *33,* 198–216.

3. The Scientific Revolution is really misnamed for there was not a single revolution, rather a series of paradigm shifts, among which the transformation in chemical thought was one of the later events. *See,* for example, Shapin, S. *The Scientific Revolution;* University of Chicago Press: Chicago, IL, 1996; Crosland, M. "Lavoisier, the Two French Revolutions and 'the Imperial Despotism of Oxygen.'" *Ambix* **1995,** *42,* 101–118.

4. Cited in Easlea, B. *Science and Sexual Oppression: Patriarchy's Confrontation with Women and Nature;* Weidenfeld & Nicolson: London, 1981; p 69. These views of the nature of women date back to early Christianity and to the classical world. *See* Bullough, V. L. "Medieval Medical and Scientific Views of Women." *Viator* **1973,** *4,* 485–501.

5. Easlea, *Science and Sexual Oppression;* p 69.

6. Cited in Easlea, *Science and Sexual Oppression;* p 70.

7. Cited in Seidel, M. A. "Poulin de la Barre's 'The Woman as Good as the Man.'" *J. Hist. Ideas* **1974,** *35,* 499–508. © The Johns Hopkins University Press. Reproduced by permission.

8. Easlea, *Science and Sexual Oppression;* p 71.

9. Meyer, G. D. *The Scientific Lady in England 1650–1760: An Account of Her Rise, with Emphasis on the Major Roles of the Telescope and Microscope;* University of California Press: Berkeley, CA, 1955; Phillips, P. *The Scientific Lady: A Social History of Women's Scientific Interests 1520–1918;* Weidenfeld & Nicholson: London, 1990.

10. Grant, D. *Margaret the First: A Biography of Margaret Cavendish, Duchess of Newcastle, 1623–1673;* Toronto University Press: Toronto, Ontario, 1957.

11. Mintz, S. I. "The Duchess of Newcastle's Visit to the Royal Society." *J. Engl. Germ. Philol.* **1952,** *51,* 168–176.

12. Cited in Easlea, *Science and Sexual Oppression;* p 69.

13. Goodman, D. "Enlightenment Salons: The Convergence of Female and Philosophic Ambitions." *Eighteenth Century Stud.* **1989,** *22,* 329–350.

14. Bodek, E. G. "Salonnieres and Bluestockings: Educated Obsolescence and Germinating Feminism." *Feminist Stud.* **1976,** *3(3/4),* 185–199. Excerpt by permission of the publisher, Feminist Studies, Inc., c/o Women's Studies Department, University of Maryland, College Park, MD 20742.

15. Iltis, C. "Madame Du Châtelet's Metaphysics and Mechanics." *Stud. Hist. Phil. Sci.* **1977,** *8,* 29–48; Edwards, S. *The Divine Mistress: A Biography of Emilie du Châtelet;* McKay: New York, 1970; Vaillot, R. *Madame Du Châtelet;* Michel: Paris, 1978; Ehrman, E. *Mme. du Châtelet;* Berg: Leamington Spa, UK, 1986.

16. Walters, R. L. "Chemistry at Cirey." In *Studies on Voltaire and the Eighteenth Century;* Besterman, T., Ed.; Institut et Musée Voltaire: Geneva, Italy, 1967; Vol. 58, pp 1807–1827.

17. Cited in Alic, M. *Hypatia's Heritage;* Beacon Press: Boston, MA, 1986; p

147. Terrall has discussed du Châtelet's strategies to balance her role as a woman and as a scientist. *See* Terrall, M. "Émilie du Châtelet and the Gendering of Science." *Hist. Sci.* **1995**, *33*, 283–310.

18. Gardiner, L. "Women in Science." In *French Women and the Age of Enlightenment;* Spencer, S. I., Ed.; Indiana University Press: Bloomington, IN, 1984; pp 181–193.

19. Taton, R. "Châtelet, Gabrielle-Emilie le Tonnelier de Breteuil, Marquise du." In *Dictionary of Scientific Biography;* Coulson, C. C., Ed.; Charles Scribner's Sons: New York, 1971; Vol. 3, pp 215–217.

20. Cited in Alic, *Hypatia's Heritage;* pp 146–147.

21. Brock, W. H. *The Fontana History of Chemistry;* Fontana: London, 1992. (Published in U.S. as *Norton History of Chemistry;* Norton: New York, 1993.)

22. "Lavoisier and the Chemical Revolution." In *The Development of Modern Chemistry;* Ihde, A. J., Ed.; Harper & Row: New York, 1964, pp 57–88; Bensaude-Vincent, B. "A Founder Myth in the History of Sciences? The Lavoisier Case." In *Functions and Uses of Disciplinary Histories;* Graham, L.; Lepenies, W.; Wiengart, Eds.; Reidel: Dortrecht, Holland, 1983; Vol. 7, pp 53–78.

23. Duveen, D. I. "Madame Lavoisier: 1758–1836." *Chymia* **1953**, *4*, 13–29.

24. Cited in Donovan, A. *Antoine Lavoisier: Science, Administration, and Revolution;* Blackwell: Oxford, UK, 1993; p 114.

25. Poirier, J.-P. "Madame Lavoisier." *Actual. Chim.* **1994**, Mar/Apr, 44–47; and Perrin, C. E. "The Lavoisier–Bucquet Collaboration: A Conjecture." *Ambix* **1989**, *36*, 5–13.

26. Bensaude-Vincent, B. "Review." *Ann. Sci.* **1995**, *52*, 197–198.

27. Vidal, M. "David Among the Moderns: Art, Science, and the Lavoisiers." *J. Hist. Ideas* **1995**, *56*, 595–623.

28. Duveen, D. I.; Klickstein, H. S. "Benjamin Franklin (1706–1790) and Antoine Laurent Lavoisier (1743–1794). Part I: Franklin and the New Chemistry." *Ann. Sci.* **1955**, *11*, 103–128; Duveen, D. I.; Klickstein, H. S. "Saltpeter, Tin, and Gunpowder: Addenda to the Correspondence of Lavoisier and Franklin." *Ann. Sci.* **1960**, *16*, 83–94.

29. Poirier, "Madame Lavoisier," pp 44–47.

30. Cited in Brody, J. "Behind Every Great Scientist . . .." *New Sci.* **1987**, *116*, 19–21.

31. Kawashima K. "Image de Mme Lavoisier à travers les trois biographies de Lavoisier." *Kagakushi* **1992**, *19*, 188–204.

32. Betham-Edwards, M. B., Ed. *Young's Travels in France During the Years 1787, 1788, 1798;* Bell: London, 1905; p 94.

33. Cited in Donovan, *Antoine Lavoisier;* p 301.

34. Beretta, M. "Chemists in the Storm: Lavoisier, Priestley and the French Revolution." *Nuncius* **1993**, *8*, 75–104.

35. Smeaton, W. A. "Madame Lavoisier, P.S. and E.I. du Pont de Nemours and the Publication of Lavoisier's 'Mémoires de chimie'." *Ambix* **1989**, *36*, 22–30.

36. French, S. J. "The du Ponts and the Lavoisiers." *J. Chem. Educ.* **1979**, *56*, 791–793.

37. Poirier, "Madame Lavoisier," p 45.

38. Cited in Smeaton, W. A. "Madame Lavoisier," p 26.

39. Cited in Smeaton, W. A. "Madame Lavoisier," p 26.

40. Thompson, M. "Causes and Circumstances of the du Pont Family's Migration." *Fr. Hist. Stud.* **1969,** *6,* 59–77.

41. Brown, S. C. *Benjamin Thomson, Count Rumford;* Massachusetts Institute of Technology Press: Cambridge, MA, 1979.

42. Cited in Duveen, D. "Madame Lavoisier: 1758–1836," p 21.

43. Cited in Duveen, D. "Madame Lavoisier: 1758–1836," p 21.

44. Cited in Sparrow, W. J. *Count Rumford of Woburn, Mass;* Thomas Y. Crowell: New York, 1964; p 250.

45. Bouchard, G. *Guyton-Morveau, chimiste et conventionnel;* Librairie Académique Perrin: Paris, 1938; Granville, A. B. "An Account of the Life and Writings of Baron Guyton de Morveau." *J. Sci. Arts* **1817,** *3,* 249–296.

46. Melhado, E. M. "Oxygen, Phlogiston, and Caloric: The Case of Guyton." *Hist. Stud. Phys. Sci.* **1983,** *13,* 311–324.

47. Smeaton, W. A. "The Contributions of P-J. Macquer, T. O. Bergman and L. B. Guyton de Morveau to the Reform of Chemical Nomenclature." *Ann. Sci.* **1954,** *10,* 87–106.

48. Rayner-Canham. G. W.; Frenette, H. "Some French Women Chemists." *Educ. Chem.* **1985,** *22,* 176–178; Ogilvie, M. *Women in Science;* Massachusetts Institute of Technology Press: Cambridge, MA, 1986; p 94.

49. Bouchard, *Guyton-Morveau,* Chapter 8. Scott Jamieson, Department of French, Sir Wilfred Grenfell College is thanked for a translation of Bouchard's biography.

50. Betham-Edwards, *Young's Travels;* pp 220, 222.

51. Smeaton, W. A. "Louis Bernard Guyton de Morveau FRS (1737–1816) and His Relations with British Scientists." *Notes Rec. R. Soc. (London)* **1967,** *22,* 113–130.

52. Grison, E.; Goupil, M.; Bret, P., Eds., *A Scientific Correspondence During the French Revolution: Louis Bernard Guyton de Morveau and Richard Kirwan, 1782–1802;* University of California at Berkeley Press: Berkeley, CA, 1994; p 7. The correspondence contains several references to the contributions of Mme. Picardet.

53. Bouchard, *Guyton-Morveau,* Chapter 21.

54. Darrow, M. H. "French Noblewomen and the New Domesticity, 1750–1850." *Feminist Stud.* **1979,** *5,* 41–65.

55. Combremont, J. de M. *Albertine Necker de Saussure;* Librairie Payot: Laussane, Switzerland, 1946.

56. Carozzi, A. V. "H.B. de Saussure, pionnier de la géologie moderne." *La Recherche* **1992,** *23,* 414–422.

57. Smeaton, W. A. "The Chemical Work of Horace Bénédict de Saussure (1740–1799), with the Text of a Letter Written to Him by Madame Lavoisier." *Ann. Sci.* **1978,** *35,* 1.

58. Combremont, J. de M. *Albertine Necker de Saussure;* Librairie Payot: Laussane, Switzerland, 1946; p 11.

59. Orr, C. C. "Albertine Necker de Saussure, The Mature Woman Author, and the Scientific Education of Women." *Women's Writing* **1995**, *2*(2), 141–153.

60. Necker de Saussure, A. A. *L' éducation progressive, ou l' étude du cour de la vie;* Sautelet: Paris, 1828.

61. The two translations into the English language were Necker de Saussure, A. A. *Progressive Education, Commencing with the Infant;* W. D. Ticknor: Boston, MA, 1835; Necker de Saussure, A. A. *Progressive Education: or, Considerations on the Course of Life;* Longman, Orme, Brown, Green, and Longman: London, 1839.

62. Necker de Saussure, A. A. *Étude de la vie des femmes;* Hauman et Cie: Brussels, Belgium, 1838.

63. Necker de Saussure, A. A. *The Study of the Life of Woman;* Lea and Blanchard: Philadelphia, PA, 1844.

64. These include Allievo, G. *Delle dottrine pedagogiche di A. Necker di Saussure;* 1884; Causse, E. *Madame Necker de Saussure et l' éducation progressive;* 1930; Masante, M. *Studio storicocritico sulla dottina pedagogica di Mme. A. Necker de Saussure;* 1920; and Maurer, P. *Albertine Adrienne Necker de Saussure als pedagogische Schriftstellerin;* 1938.

65. Cited in Alic, *Hypatia's Heritage;* p 93.

66. Darrow, "French Noblewomen and the New Domesticity, 1750–1850."

67. Cited in Patterson, E. C. *Mary Somerville and the Cultivation of Science, 1815–1840;* Nijhoff: The Hague, Netherlands, 1983; p 22. Mme. Biot was Francoise Gabrielle Brisson. *See* Crosland, M. *The Society of Arcueil: A View of French Science at the Time of Napoleon I;* Harvard University Press: Cambridge, MA, 1967; p 96.

Chapter 3

1. Ainley, M. G. "D'assistantes anonymes à chercheures scientifiques: une rétrospective sur la place des femmes en sciences." *Cah. Rech. Sociolog.* **1986**, *4*, 55–71.

2. Kauffman, G. B. "Svante August Arrhenius, Swedish Pioneer in Physical Chemistry." *J. Chem. Educ.* **1988**, *65*, 437–438.

3. Von Hofmann, A. W. "Zur Erinnerung an Friedrich Wöhler." *Berichte* **1882**, *15*, 3127–3290.

4. Trescott, M. M. "Julia B. Hall and Aluminum." *J. Chem. Educ.* **1977**, *54*, 24–25; and Trescott, M. M. "Julia B. Hall and Aluminum." In *Dynamos and Virgins Revisited: Women and Technological Change in History;* Moore, M. M., Ed.; Scarecrow: Metuchen, NJ, 1979; pp 149–179. Trescott claims that Julia Hall was essentially co-inventor of the aluminum extraction process; however, in a more recent (unpublished) study, N.C. Craig of Oberlin College has shown that Julia acted more as observer and assistant.

5. Stock, A. "Henri Moissan." *Berichte* (Band 4) **1907**, *40*, 5099–5130.

6. Houlihan, S.; Wotiz, J. H. "Women in Chemistry Before 1900." *J. Chem. Educ.* **1975**, *52*, 362–364.

7. Houlihan, S.; Wotiz, J. H. "Women in Chemistry Before 1900," p 363.

8. Kohlstedt, S. G. "In from the Periphery: American Women in Science, 1830–1880." *Signs* **1978**, *4(1)*, 81–96.

9. Phillips, P. *The Scientific Lady: A Social History of Women's Scientific Interests 1520–1918;* St. Martin's Press: New York, 1990.

10. Morrison-Low, A. D. "Women in the 19th Century Scientific Instrument Trade in Britain." *Bull. Sci. Instrum. Soc.* **1991**, *28*, 7–11.

11. Quednau, W. *Maria Sibylla Merian: Der Lebensweg einer grossen Küntslerin und Forscherin;* Mohn: Gütersloh, 1966; Stearn, W. T. "Maria Sibylla Merian (1647–1717) as a Botanical Artist." *Taxon* **1982**, *31*, 529–534; Davis, N. Z. *Women on the Margins: Three Seventeenth Century Lives;* Harvard University Press: Cambridge, MA, 1995; Alic, M. "The New Naturalists." In *Hypatia's Heritage;* Beacon Press: Boston, MA, 1986; pp 108–118.

12. Wilson, J. H. "Dancing Dogs of the Colonial Period: Women Scientists." *Early Am. Lit.* **1972/73**, *7*, 225–235.

13. Dobson, A. K.; Bracher, K. "A Historical Introduction to Women in Astronomy." *Mercury* **1992**, *21(1)*, 4–15; Aldrich Kidwell, P. "Women Astronomers in Britain, 1780–1930." *Isis* **1984**, 75, 534–546; Lankford J.; Slavings, R. L. "Gender and Science: Women in American Astronomy, 1859–1940." *Phys. Today* **1990**, *43(3)*, 58–65.

14. Elder, E. S. "Women in Early Geology." *J. Geol. Educ.* **1982**, *30(5)*, 287–293.

15. Bucciarelli, L. L.; Divorsky, N. *Sophie Germain;* D. Reidel: Dordrecht, Holland, 1980; Truesdale, C. "Sophie Germain: Fame Earned by Stubborn Error." *Boll. Storia Sci. Matemat.* **1991**, *11(2)*, 3–24; Dehan-Dalmedico, A. "Sophie Germain." *Sci. Am.* **1991**, *265(6)*, December 116–120; Sampson, J.H. "Sophie Germain and the Theory of Numbers." *Arch. Hist. Exact Sci.* **1990**, *41*, 157–161.

16. Weldon, W., cited in Golinski, J. *Science as Public Culture: Chemistry and Enlightenment in Britain, 1760–1820;* Cambridge University Press: Cambridge, UK, 1992; p 160.

17. For other accounts of Elizabeth Fulhame, *see* Davenport, D. A.; Ireland, K. M. "The Ingenious, Lively and Celebrated Mrs. Fulhame and the Dyer's Hand." *Bull. Hist. Chem.* **1989**, *5*, 37–42; Rayner-Canham, G. W. "Two British Women Chemists." *Educ. Chem.* **1983**, *20*, 140–141.

18. Fulhame, Mrs. *An Essay on Combustion with a View to a New Art of Dying and Painting, Wherein the Phlogistic and Antiphlogistic Hypotheses Are Proved Erroneous;* J. Cooper: London, 1794.

19. Fulhame, Mrs. *An Essay on Combustion,* p xix.

20. Schaaf, L. J. "The First Fifty Years of British Photography: 1794–1844." In *Technology and Art: The Birth and Early Years of Photography;* Pritchard, M., Ed.; Royal Photographic Society Historical Group: Bath, UK, 1990, pp 10–12; *see also,* Laidler, K. J. "The Story of Photochemical Imaging." *Chem News* **1992**, *13 (April)*, 8.

21. Laidler, K. J. "The Development of Theories of Catalysis." *Arch. Hist. Exact Sci.* **1986**, *35*, 345–374.

22. Hening, G. R. "Surface Reactions of Single Crystals of Graphite." *J. Chim. Phys.* **1961**, *58*, 12–19.

23. Anon. "An Essay on *Combustion, by a Lady!*" *Gentleman's Mag.* **1795**, *65(June)*, 501.

24. Coindet, J. F. *Ann. Chim.* **1797**, *26* (series 1), 58.

25. Cited in Davenport and Ireland, "The Ingenious, Lively and Celebrated Mrs. Fulhame and the Dyer's Hand," p 37.

26. Cited in Wheeler T. S.; Partington, J. R. *The Life and Work of William Higgins, Chemist (1763–1825);* Pergamon Press: Oxford, UK, 1960; p 110.

27. Woodhouse, J. "An Answer to Dr. Joseph Priestley's Considerations on the Doctrine of Phlogiston, and the Decomposition of Water; Founded upon Demonstrative Experiments." *Trans. Am. Phil. Soc.* **1799**, *4*, 465.

28. Miles, W. "Early American Chemical Societies." *Chymia* **1950**, *3*, 95–113.

29. Cited in Smith, E. F. *Chemistry in America: Chapters from the History of the Science in the United States;* Reprint, Arno Press: New York, 1972; p 35.

30. Fulhame, Mrs. *Versuche über die Wiederherstellung der Metalle durch Wasserstoffgas, Phosphor, Schwefel, Schwefelleber, geschwefeltes Wasserstoffgas, gephosphortes Wasserstoffgas,* translated by A. G. W. Lenten; Göttingen, Germany, 1798.

31. Fulhame, Mrs. *An Essay on Combustion;* Humphries: Philadelphia, PA, 1810.

32. Cited in Davenport and Ireland, "The Ingenious, Lively and Celebrated Mrs. Fulhame and the Dyer's Hand," p 41.

33. Mellor, J. W. "History of the Water Problem (Mrs. Fulhame's Theory of Catalysis)." *J. Phys. Chem.* **1903**, *7*, 557–567.

34. Alarotti, F. *Il Neutonianismo per la dame;* 1737; Carter, E. *Sir Isaac Newton's Philosophy Explain'd for the Use of the Ladies;* 1739.

35. Rossiter, M. W. "Women and the History of Scientific Communication." *J. Library Hist.* **1986**, *21*, 39–59.

36. Perl, T. "The Ladies' Diary or Woman's Almanack, 1704–1841." *Hist. Math.* **1979**, *6*, 36–53; Perl, T. H. "Women and Mathematics in the Late 18th Century: A Query." *Hist. Math.* **1980**, *7*, 188–189; Wallis R.; Wallis, P. "Female Philomaths." *Hist. Math.* **1980**, *7*, 57–64.

37. Armstrong, E. V. "Jane Marcet and Her *Conversations on Chemistry*." *J. Chem. Educ.* **1938**, *15*, 53–57. See also Derrick, E. M. "What Can a Nineteeth Century Chemistry Textbook Teach Twentieth Century Chemists?." *J. Chem. Educ.* **1985**, *62(9)*, 749–751; Knight, D. "Accomplishment or Dogma: Chemistry in the Introductory Works of Jane Marcet and Samuel Parkes." *Ambix* **1986**, *33(2/3)*, 94–98; and "Jane Haldimand Marcet (1769–1858)." In Thomson, D. L. *Adam Smith's Daughters;* Exposition Press: New York, 1973; pp 9–28.

38. Smeaton, W. A. "French Scientists in the Shadow of the Guillotine: The Death Roll of 1792–1794." *Endeavour* **1993**, *17*, 60–63.

39. Brock, W. H. *The Norton History of Chemistry;* Norton: New York, 1993.

40. Golinski, J. *Science as Public Culture: Chemistry and Enlightenment in Britain, 1760–1820;* Cambridge University Press: Cambridge, UK, 1992; p 194.

41. McDermid, J. "Conservative Feminism and Female Education in the Eighteenth Century." *Hist. Educ.* **1989**, *18*, 309–322.

42. The manuscript was edited by Alexander Marcet's friend, the London physician, John Yelloy. See Crellin, J. K. "The Story Behind the Story: Mrs Marcet's *Conversations on Chemistry*." *J. Chem. Educ.* **1979**, *56(7)*, 459–460.

43. Anon., *Conversations on Chemistry: In Which the Elements of That Science Are Familiarly Explained and Illustrated by Experiments,* 1st ed; Longman, Hurst, Rees and Orme: London, 1806.

44. Shteir, A. B. "Botanical Dialogues: Maria Jacson and Women's Popular Science Writing in England." *Eighteenth-Century Stud.* **1990,** *23(3),* 301–317.

45. Anon, *Conversations on Chemistry.*

46. Anon, *Conversations on Chemistry.*

47. Lindee, S. M. "The American Career of Jane Marcet's *Conversations on Chemistry,* 1806–1853." *Isis* **1991,** *82,* 8–23.

48. Warner, D. J. "Science Education for Women in Antebellum America." *Isis* **1978,** *69,* 58–67.

49. Lindee, "The American Career of Jane Marcet's *Conversations on Chemistry,* 1806–1853."

50. Miles, W. "Books on Chemistry Printed in the United States, 1755–1900: A Study of Their Origin." *Libr. Chron. (Philadelphia)* **1952,** *18,* 51–62.

51. Jacques, J. "Une Chemiste qui Avait de la Conversation." *Nouv. J. Chim.* **1986,** *10,* 209–211.

52. Cole, W. A. *Chemical Literature 1700–1860: A Bibliography with Annotations, Detailed Descriptions, Comparisons and Locations;* Mansell: London, 1988, p 420.

53. James, A. J. L. "Michael Faraday—the Chemist." *Educ. Chem.* **1991,** *28,* 128–130.

54. Michael Faraday to Auguste de la Rive, cited in Armstrong, "Jane Marcet and Her *Conversations on Chemistry,*" p.56.

55. Cited in Armstrong, "Jane Marcet and Her *Conversations on Chemistry,*" p 56.

56. Jane Marcet to Michael Faraday, 24 November 1845, cited in Armstrong, "Jane Marcet and Her *Conversations on Chemistry,*" p 56.

57. Edgeworth, M. *Letters for Literary Ladies;* London, 1795; reprint; Garland: New York, 1974; p 66.

58. Benjamin, M. "Elbow Room: Women Writers on Science, 1790–1840." In *Science and Sensibility: Gender and Scientific Enquiry, 1780–1945;* Benjamin, M., Ed.; Blackwell: Oxford, UK, 1991; pp 27–59.

59. Maria Edgeworth to Harriet Beaufort, 5 January 1822, cited in Colvin, C., Ed.; *Maria Edgeworth, Letters from England 1813–1844;* Clarendon Press: Oxford, UK, 1971, p 308.

60. Fussell, G. E. "Some Lady Botanists of the Nineteenth Century: V. Jane Marcet." *The Gard. Chron.* **1951,** (December 22), *130* (series 3), 238; Shtier, A. B. *Cultivating Women, Cultivating Science: Flora's Daughters and Botany in England 1760–1860;* Johns Hopkins University Press: Baltimore, MD, 1996; pp 99–102.

61. Cited in Fussell, "Some Lady Botanists of the Nineteenth Century: V. Jane Marcet."

62. The life of Michael is summarized from a memorial volume published by request of the Michael family: Michael, H. A. *Studies in Plant and Organic Chemistry and Literary Papers;* Riverside Press: Cambridge, MA, 1907; and

from Tarbell, A. T.; Tarbell, D. S. "Helen Abbott Michael: Pioneer in Plant Chemistry." *J. Chem. Educ.* **1982**, *59*, 548–549.

63. Helmholtz, H. L. F. *Handbuch der physiologischen Optik;* Leipzig, 1867.

64. Michael, *Studies in Plant and Organic Chemistry and Literary Papers*, p 9.

65. Alsop, G. F. *History of the Woman's Medical College, Philadelphia, Pennsylvania, 1850–1950;* Lippincott: Philadelphia, PA, 1950.

66. Michael, *Studies in Plant and Organic Chemistry and Literary Papers*, p 9.

67. Heiser, C. B. "Taxonomy." In *A Short History of Botany in the United States*; Ewan, J., Ed.; Hefner: New York, 1969; p 114.

68. Abbott, H. C. de S. "Certain Chemical Constituents of Plants Considered in Relation to Their Morphology and Evolution." *Bot. Gaz.* **1886**, *11*, 270–272.

69. Abbott, H. "The Chemical Basis of Plant Forms." *J. Franklin Inst.* **1887**, *124*, 161–185; Abbott, H. "Comparative Chemistry of Higher and Lower Plants." *Am. Nat.* **1887**, *21*, 719–800.

70. Cited in Tarbell and Tarbell, "Helen Abbott Michael: Pioneer in Plant Chemistry," p 548.

71. Though individual professors welcomed Michael, they were unwilling to risk official wrath by admitting her to their laboratories. *See* Albisetti, J. C. *Schooling German Girls and Women: Secondary and Higher Education in the 19th Century;* Princeton University Press: Princeton, NJ, 1988; p 139.

72. The most complete accounts of Pockels's life and work are given by Giles C. H.; Forrester, S. D. "The Origins of the Surface Film Balance: Studies in the Early History of Surface Chemistry, part 3." *Chem. Ind.* **1971**, 9 January, 43–53; and Beisswanger, G. "Agnes Pockels (1862–1935) und die Oberflächenchemie." *Chem. Unserer Zeit* **1991**, *25*, 97–101. The former article has been summarized in Derrick, M. E. "Agnes Pockels, 1862–1935." *J. Chem. Educ.* **1982**, *59*, 1030–1031.

73. Cited in Giles and Forrester, "The Origins of the Surface Film Balance," p 48.

74. Cited in Giles and Forrester, "The Origins of the Surface Film Balance," p 46.

75. Cited in Giles and Forrester, "The Origins of the Surface Film Balance," p 47.

76. Cited in Giles and Forrester, "The Origins of the Surface Film Balance," p 47.

77. A. Pockels to Lord Rayleigh, 10 January 1891, cited in Giles and Forrester, "The Origins of the Surface Film Balance," p 43.

78. A. Pockels to Lord Rayleigh, 1 February 1891, cited in Giles and Forrester, "The Origins of the Surface Film Balance," p 44.

79. Lord Rayleigh to Sir J. N. Lockyer (editor of *Nature*), 2 March 1891, cited in Giles and Forrester, "The Origins of the Surface Film Balance," p 49.

80. Rayleigh Lord and Pockels, (Miss) A. "Letters to the Editor: Surface Tension." *Nature (London)* **1891**, *43*, 437–439.

81. Pockels, (Miss) A. "On the Relative Contamination of the Water Surface by Equal Quantities of Different Substances." *Nature (London)* **1892**, *46*, 418–419.

82. Most of her papers are listed in a review of her work published on her 70th birthday: Ostwald, W. "Die Arbeiten von Agnes Pockels über Grenz-schichten und Filme." *Kolloid-Z.* **1932**, *58*, 1–8.

83. Her last English-language paper was Pockels, A. "The Measurement of Surface Tension with the Balance." *Science (Washington, D.C.)* **1926**, *64*, 304.

84. Cited in Giles and Forrester, "The Origins of the Surface Film Balance," p 50.

85. Cited in Giles and Forrester, "The Origins of the Surface Film Balance," p 50.

86. Beisswanger (ref. 72) could not trace any record as to the identity of Laura R. Leonard or the reason for naming the prize after her.

87. *See,* for example, Freifelder, D. *Principles of Physical Chemistry with Applications to the Biological Sciences,* 2nd. ed.; Jones and Bartlett: Boston, MA, 1985; pp 430–431, 434–436; Adamson, A. W. *Physical Chemistry of Surfaces,* 5th. ed.; Wiley-Interscience: New York, 1990; pp 181–183.

88. Alic, M. "The Popularisation and Professionalisation of Science." *Hypatia's Heritage: A History of Women in Science from Antiquity Through the Nineteenth Century;* Beacon Press: Boston, MA, 1986; p 180.

89. Patterson, E. C. *Mary Somerville and the Cultivation of Science, 1815–1840;* Nijhof/Kluwer: Boston, MA, 1983.

90. James, F. *The Correspondence of Michael Faraday, volume 2;* Institution of Electrical Engineers: London, 1993; letters 821 and 824.

91. Somerville, M. *On the Connexion of the Physical Sciences;* J. Murray: London, 1846 (reprint edition, Arno Press: New York, 1975).

Chapter 4

1. Lange, H. *Higher Education of Women in Europe,* translated by L.B. Klemm; Appleton: New York: 1890. There are two chronicled accounts of women trying to gain access to higher education in earlier times. *See* Shank, M. H. "A Female University Student in Late Medieval Krakow." *Signs* **1987**, *12*, 373–380; Schiebinger, L. "Maria Winkelmann at the Berlin Academy: A Turning Point for Women in Science." *Isis* **1987**, *78*, 174–200.

2. Manthorpe, C. "Science or Domestic Science? The Struggle To Define an Appropriate Science Education for Girls in Early 20th Century England." *Hist. Educ.* **1986**, *15*, 195–213; Conway, J. K. "Perspectives on the History of Women's Education in the United States." *Hist. Educ. Quart.* **1974** (Spring), *14*, 1–12; Tolley, K. "Science for Ladies, Classics for Gentlemen: A Comparative Analysis of Scientific Subjects in the Curricula of Boys' and Girls' Secondary Schools in the United States, 1794–1850." *Hist. Educ. Quart.* **1996** (Summer), *36*, 128–153.

3. Cited in Stetson, D. McB. *Women's Rights in France;* Greenwood Press: Westport, CT, 1987; p 112.

4. Moses, C. G. *French Feminism in the Nineteenth Century;* State University of New York Press: Albany, NY, 1990; p 175.

5. Stetson, D. McB. *Women's Rights in France;* Greenwood Press: Westport, CT, 1987; p 113.

6. Frevert, U. *Women in German History;* Berg: Oxford, UK, 1989; p 124. *See also* Rayner-Canham, M. F.; Rayner-Canham, G. W. "Elsa Neumann (1872–1902): A Tragic Loss to Science." *Bull. Hist. Chem.*, submitted.

7. MacLeod, R.; Moseley, R. "Fathers and Daughters: Reflections on Women, Science, and Victorian Cambridge." *Hist. Educ.* **1979,** *8,* 321–333.

8. Becker, L. E. "On the Study of Science by Women." *Contemp. Rev.* **1869,** *10,* 386–404.

9. Wein, R. "Women's Colleges and Domesticity, 1875–1918." *Hist. Educ. Quart.* **1974,** *14,* 31–48.

10. From Sewell's *Women and the Times We Live In;* cited in Purvis, J. *A History of Women's Education in England;* Open University Press: Milton Keynes, UK, 1991; pp 111–112.

11. Cited in Conway, J. "Stereotypes of Femininity in a Theory of Sexual Evolution." In *Suffer and Be Still: Women in the Victorian Age;* Vicinus, M., Ed.; Indiana University Press: Bloomington, IN, 1972; p 141.

12. Zschoche, S. "Dr. Clarke Revisited: Science, True Womanhood, and Female Collegiate Education." *Hist. Educ. Quart.* **1989,** *29,* 545–569; Alaya, F. "Victorian Science and the 'Genius' of Woman." *J. Hist. Ideas* **1977,** *38,* 261–280.

13. Cited in Fedigan, L. M. "The Changing Role of Women in Models of Human Evolution." In *Inventing Women: Science, Technology, and Gender;* Kirkup, G.; Keller, L. S., Eds.; Polity Press: Cambridge, UK, 1992; pp 103–122.

14. Hubbard, R. "Feminism in Academia: Its Problematic and Problems." In *Expanding the Role of Women in the Sciences;* Briscoe, A. M.; Pfafflin S. M., Eds.; New York Academy of Sciences: New York, 1979; p 251. *See also* Burstyn, J. N. "Education and Sex: The Medical Case Against Higher Education for Women in England, 1870–1900." *Proc. Am. Phil. Soc.* **1973,** *117,* 79–89.

15. *Sex and Education: A Reply to Dr. Clarke's "Sex in Education."* Howe, J. W., Ed.; Roberts: Boston, MA, 1874, reprinted by Arno Press, New York, 1972. Julia Ward Howe (1819–1910) was a founder and prime mover of numerous suffrage, relief, and peace organizations. It was her belief in the necessity of higher education of young women that caused her to rally the progressive forces against Clarke's philosophies.

16. Newman, L. M. "The Evils of Education (1870–1900)." In *Men's Ideas/ Women's Realities: Popular Science, 1870–1915;* Newman, L. M., Ed.; Pergamon: New York, 1984; pp 54–104.

17. Solomon, B. Miller. *In the Company of Educated Women;* Yale University Press: New Haven, CT, 1985; Gordon, L. D. *Gender and Higher Education in the Progressive Era;* Yale University Press: New Haven, CT, 1990; and Boas, L. S. *Women's Education Begins: The Rise of the Women's Colleges;* reprint Arno Press: New York, 1971.

18. Wein, R. "Women's Colleges and Domesticity, 1875–1918." *Hist. Educ. Quart.* **1974** (Spring), *14,* 31–47.

19. Whitney, M. W. "Scientific Study and Work for Women." *Education* **1882,** *3,* 58–69.

20. Hogeland, R. W. "Co-Education of the Sexes at Oberlin College: A Study of Social Ideas in Mid-Nineteenth Century America." *J. Social Hist.* **1972– 1973** (Winter), *6*, 160–176.

21. Rury, J.; Harper, G. "The Trouble with Coeducation: Mann and Women at Antioch, 1853–1860." *Hist. Educ.* **1986**, *26*, 481–502.

22. Bishop, M. *A History of Cornell;* Cornell University Press: Ithaca, NY, 1962; pp 143–152.

23. Durbin, N. E.; Kent, L. "Postsecondary Education of White Women in 1900." *Sociology Educ.* **1989**, *62*, 1–13.

24. Newcomer, M. *A Century of Higher Education for American Women;* Harper: New York, 1959, p 37; Graham, P. A. "Expansion and Exclusion: A History of Women in American College Education." *Signs* **1978** (Summer), *3*, 759–773.

25. Kamm, J. *Hope Deferred: Girl's Education in English History;* Methuen: London, 1965.

26. MacLeod, R.; Moseley, R. "Fathers and Daughters: Reflections on Women, Science and Victorian Cambridge." *Hist. Educ.* **1979**, *8*, 321–333; McWilliams-Tullberg, R. *Women at Cambridge;* Gollanz: London, 1975.

27. Bernstein, G.; Bernstein, L. "Attitudes Towards Women's Education in Germany, 1870–1914." *Int. J. Women's Stud.*, **1979**, *2(5)*, 473–488.

28. Albisetti, J. C. *Schooling German Girls and Women: Secondary and Higher Education in the 19th Century;* Princeton University Press: Princeton, NJ, 1988; pp 122–135.

29. Bridges, F. "Coeducation in Swiss Universities." *Pop. Sci. Mon.* **1890**, *38*, 524–530.

30. Bernal, I.; Kauffman, G. B. "The Spontaneous Resolution of cis-Bis(ethylenediamine)dinitrocobalt(III) salts: Alfred Werner's Overlooked Opportunity." *J. Chem. Educ.* **1987**, *64*, 604–610.

31. Johanson, C. *Women's Struggle for Higher Education in Russia, 1855–1900;* McGill-Queen's University Press: Kingston, ON, 1987.

32. Edmondson, L. H. *Feminism in Russia, 1900–1917;* Stanford University Press: Stanford, CA, 1984.

33. Koblitz, A. H. "Science, Women, and the Russian Intelligencia." *Isis* **1988**, *79*, 208–226.

34. Hayes, A. "Can a Poor Girl Go to College?" *North Am. Rev.* **1891**, *152*, 624–631.

35. Solomon, B. M. *In the Company of Educated Women;* Yale University Press: New Haven, CT, 1985; p 63.

36. McGrayne, S. B. *Nobel Prize Women in Science;* Birch Lane Press: New York, 1992; p 6.

37. Sulloway has argued that birth order is of vital importance in determining traits; in particular, he argues that the first-born child has unique behavioral characteristics. *See* Sulloway, F. J. *Born to Rebel: Birth Order, Family Dynamics and Creative Lives;* Random House of Canada: Toronto, ON, 1996.

38. Fingard, J. "College, Career, and Community: Dalhousie Coeds, 1881–

1921." In *Youth, University and Canadian Society;* Axelrod, P.; Reid, J. G., Eds.; McGill-Queen's University Press: Montreal, PQ, 1989; p 26.

39. Thomas, M. C. "Present Tendencies in Women's College and University Education." *Educ. Rev.* 1908, *35,* 64–85; reproduced in *And Jill Came Tumbling After—Sexism in American Education;* Stacey J. et al., Eds.; Laurel: New York, 1974; p 276. *See also* Horowitz, H. L. *The Power and Passion of M. Cary Thomas;* A. A. Knopf: New York, 1994.

40. Vicinus, M. *Independent Women: Work and Community for Single Women 1850–1920;* University of Chicago Press: Chicago, IL, 1985; p 138.

41. Cited in Vicinus, M. "One Life to Stand Beside Me: Emotional Conflicts in First-Generation College Women in England." *Feminist Stud.* 1982, *8,* 603–628. Excerpt by permission of the publisher, Feminist Studies, Inc., c/o Women's Studies Department, University of Maryland, College Park, MD 20742.

42. Derick, C. M. "In the 80's." *Old McGill* 1927, 200.

43. Purvis, *A History of Women's Education in England,* p.117; *see also* LaPierre, J. "The Academic Life of Co-eds, 1880–1900," *Hist. Stud. Educ.* 1990, *2,* 225–245.

44. W.H.B. "Mrs. G.P. Bidder." *Nature (London)* 1932, *130,* 689–690. Reprinted with permission, © 1932 Macmillan Magazines Limited. Mrs. G. P. Bidder was the married name of Marion Greenwood.

45. Ball, M. D. "Newnham Scientists." In *A Newnham Anthology;* Phillips, A., Ed.; Cambridge University Press: Cambridge, UK, 1979; p 77.

46. Cited in Frevert, *Women in German History,* pp 121–122.

47. Harrison, J. F. C. *Late Victorian Britain 1875–1901;* Fontana Press: London, 1990; p 171.

48. Rossiter, M. W. "'Womens Work' in Science, 1880–1910." *Isis* 1980, *71,* 381–398; also in Rossiter, M. W. *Women Scientists in America,* p 52.

49. Antler, J. "'After College What?': New Graduates and the Family Claim." *Am. Quart.* 1980 (Fall), *32,* 409–434.

50. Cookingham, M. E. "Bluestockings, Spinsters, and Pedagogues: Women College Graduates, 1865–1910." *Pop. Stud.* 1984, *38,* 349–364.

51. Rossiter, M. W. *Women Scientists in America,* p 73.

52. Alsop, G. F. "Rachel Bodley, 1831–1888." *J. Am. Med. Women's Assoc.* 1949, *4,* 534–536.

53. Cited in Kauffman, G. B. "The Misogynist Dinner of the American Chemical Society." *J. Coll. Sci. Teaching* 1983, *12,* 381–382. The booklet was compiled and published in about 1880 from notes taken at the event by Henry Morton.

54. Mason, J. "A Forty Years' War." *Chem. Br.* 1991, *27,* 233–238. Mason has also described the exclusion of women from the Royal Society. *See* Mason, J. "The Admission of the First Women to the Royal Society of London." *Notes Records R. Soc. London* 1992, *46,* 279–300.

55. Cited by Mason, "A Forty Years' War," p 237.

56. "Editorial: Women and the Fellowship of the Chemical Society." *Nature (London)* 1908, *78,* 226. Reprinted with permission, © 1908 Macmillan Magazines Limited.

57. Whiteley, M. A. "Ida Smedley MacLean. 1877–1944." *J. Chem. Soc.* 1946, 65–67.

58. Creese, M. R. S.; Creese, T. M. "Ellen Henrietta Swallow Richards (1842–1911)." In *Women in Chemistry and Physics: A Biobibliographic Sourcebook*; Grinstein, L. S.; Rose, R. K.; Rafailovich, M. H., Eds.; Greenwood Press: Westport, CT, 1993; pp 515–525; Clarke, R. *Ellen Swallow: The Woman Who Founded Ecology*; Follett: Chicago, IL, 1973; Hunt, C. L. *The Life of Ellen H. Richards;* Whitcomb and Barrows: Boston, MA, 1912.·

59. Cited in Clarke, R. *Ellen Swallow: The Woman Who Founded Ecology;* Follett: Chicago, IL, 1973; pp 13–14.

60. Cited in Bernard, J. *Academic Women;* Meridian: New York, 1984; p 9.

61. Cited in Hunt, *The Life of Ellen H. Richards,* p 91.

62. Cited in Rosen, G. "Ellen H. Richards (1824–1911): Sanitary Chemist and Pioneer of Professional Equality for Women in Health Sciences." *Am. J. Public Health* **1974,** *64,* 312–323.

63. Richards, E. H.; Woodman, A. G. *Air, Water and Food from a Sanitary Standpoint;* J. Wiley: New York, 1900.

64. Richards, E. H. *Food Materials and Their Adulterations;* Estes and Lauriat: Boston, MA, 1886.

65. Richards, E. H. *The Chemistry of Cooking and Cleaning: A Manual for Housekeepers;* Estes and Lauriat: Boston, MA, 1881.

66. Cravens, H. "Establishing the Science of Nutrition at the USDA: Ellen Swallow Richards and Her Allies." *Agric. Hist.* **1990,** *64(2),* 122–133.

67. The Rumford Kitchen was named after Count Rumford, Benjamin Thomson, who had devised the modern cooking stove. *See* Brown, S. C. *Benjamin Thomson, Count Rumford;* Massachusetts Institute of Technology Press: Cambridge, MA, 1979. As discussed in Chapter 2 of this book, Thomson was also the second husband of Marie Lavoisier.

68. Richards, E. H. *Euthenics, the Science of Controllable Environment; a Plea for Better Living Conditions as the First Step Toward Higher Human Efficiency;* Whitcomb and Barrows: Boston, MA, 1912. *See also* Weigley, E. S. "It Might Have Been Euthenics: The Lake Placid Conferences and the Home Economics Movement." *Am. Quart.* **1974,** *26,* 79–96.

69. Ogilvie, M. B. *Women in Science, Antiquity Through the Nineteenth Century: A Biographical Dictionary with Annotated Bibliography;* Massachusetts Institute of Technology Press: Cambridge, MA, 1986; pp 149–152.

70. Whiteley, "Ida Smedley MacLean. 1877–1944," pp 65–67.

71. Sloan, J. B. "The Founding of the Naples Table Association for Promoting Scientific Research by Women, 1897." *Signs* **1978,** *4,* 208–216.

72. Tarbell, A. T.; Tarbell, D. S. "Dr. Rachel Lloyd (1839–1900): American Chemist." *J. Chem. Educ.* **1982,** *59,* 743–744; Creese, M. R. S.; Creese, T. M. "Rachel Lloyd: Early Nebraska Chemist." *Bull. Hist. Chem.* **1995,** *17/18,* 9–14.

73. Cited in Creese, M. R. S.; Creese, T. M. "Rachel Lloyd: Early Nebraska Chemist," p 9.

74. Maberly, C. F. "Obituary." *J. Am. Chem. Soc. Proc.* **1901,** *33,* 84.

75. Creese, M. R. S.; Creese, T. M. "Laura Alberta Linton (1853–1915): An American Chemist." *Bull. Hist. Chem.* **1990,** *8,* 15–18.

76. Peckham, S. F.; Hall, C. W. "On Lintonite and Other Forms of Thomsonite: A Preliminary Notice of the Zeolites in the Vicinity of Grand Marais,

Cook County, Minnesota." *Am. J. Sci.* **1880,** *19,* 122–130; *see also,* Dahlberg, C. "Laura A. Linton and Lintonite," *Minn. Hist.* **1962,** *38(1),* 21–23.

77. Dahlberg, J. C. "A Woman to Remember." *Lapidary J.* **1976** (Oct), *29,* 1732–1736.

78. Creese, M. R. S. "British Women of the Nineteenth and Early Twentieth Centuries Who Contributed to Research in the Chemical Sciences." *Br. J. Hist. Sci.* **1991,** *24,* 275–305. This quotation is reproduced with permission of the Council of the British Society for the History of Science.

79. E. W., "Ida Freund." *Girton Rev.* **1914,** 9–13.

80. Wilson, H. "Miss Freund." In *A Newnham Anthology;* Phillips, A., Ed.; Cambridge University Press: Cambridge, UK, 1979; pp 71–72.

81. Ball, M. D. "Newnham Scientists." In *A Newnham Anthology;* Phillips, A., Ed.; Cambridge University Press: Cambridge, UK, 1979; p 76.

82. Berry, A. J.; Moelwyn-Hughes, E. A. "Chemistry at Cambridge from 1901 to 1910." *Proc. Chem. Soc.* **1963,** 357–363.

83. Freund, I. *The Study of Chemical Composition. An Account of Its Method and Historical Development, with Illustrative Quotations;* The University Press: Cambridge, UK, 1904 (reprint edition with a biographical essay, Dover Publications: New York, 1968).

84. Quoted in Gardiner, M. I. "In Memoriam—Ida Freund." *Newnham College Roll Lett.* **1914,** 34–38.

85. Freund, I. *The Study of Chemical Composition. An Account of Its Method and Historical Development;* reprint edition, Dover Publications: New York, 1968.

86. Freund, I. *The Experimental Basis of Chemistry; Suggestions for a Series of Experiments Illustrative of the Fundamental Principles of Chemistry;* The University Press: Cambridge, UK, 1904.

87. Cited in Mason, "A Forty Years' War," p 234.

88. Steinberg, C. "Yulya Vsevolodovna Lermontova (1846–1919)." *J. Chem. Educ.* **1983,** *60,* 757–758.

89. Sonya Sophia Kovalevskaia (1850–1891) was an outstanding mathematician. It was the custom for Russian women of the time who desired an education to contract a marriage of convenience with a man who was also planning to attend a foreign university. The details of her particular adventures are described in Koblitz, A. H. *A Convergence of Lives: Sofia Kovalevskaia, Scientist, Writer, Revolutionary;* Birkhauser: Cambridge, MA, 1983; her personal life is the focus of Kennedy, D. H. *Little Sparrow: A Portrait of Sophia Kovalevsky;* Ohio University Press: Athens, OH, 1983; her contributions to mathematics are described in Cooke, R. *The Mathematics of Sonya Kovalevskaya;* Springer-Verlag: New York, 1984; Keen, L., Ed. *The Legacy of Sonya Kovalevskaya;* American Mathematical Society: Providence, RI, 1987; Koblitz, A. "Sofia Kovalevskaia and the Mathematical Community." *Mat. Intell.* **1984,** *6(1),* 20–29.

90. These quotes are cited in Stillman, B. "Sofya Kovalevskaya: Growing Up in the Sixties." *Russ. Lit. Triquart.* **1974** (Spring), *9,* 276–302.

91. Koblitz, A. H. "Career and Home Life in the 1880s: The Choices of Mathematician Sofia Kovalevskaia." In *Uneasy Careers and Intimate Lives: Women*

in Science 1789–1979; Abir-Am, P. G.; Outram, D., Eds.; Rutgers University Press: New Brunswick, NJ, 1987; p 178.

92. Cited in Kennedy, D. H. *Little Sparrow,* pp 144–146. *See also* Koblitz, A. H. *A Convergence of Lives,* p 102.

93. Stillman, B. "Sofya Kovalevskaya: Growing Up in the Sixties," p. 285.

94. Cited in Koblitz, A. H. *A Convergence of Lives: Sophia Kovalevskaia, Scientist, Writer, Revolutionary;* Birkhauser: Cambridge, MA, 1983; p 127. This book documents the intertwining of the lives of Lermontova and Kovalevskaia.

95. Cited in Koblitz, A. H. *A Convergence of Lives,* p 145.

96. Elder, E. S.; Lazzerini, S. "The Deadly Outcome of Chance—Vera Estaf'evna Bogdanovskaia." *J. Chem. Educ.* **1979,** *56,* 251–252. The second woman chemist to die in an accident was Elsa Neumann. *See* Rayner-Canham, M. F.; Rayner-Canham, G. W. "Elsa Neumann (1872–1902): A Tragic Loss to Science."

97. "Notes." *Nature (London)* **1897,** *56,* 132.

98. Gier, T. E. "HCP, A Unique Phosphorus Compound." *J. Am. Chem. Soc.* **1961,** *83,* 1769–1770.

99. Cited in Elder and Lazzerini, "The Deadly Outcome of Chance—Vera Estaf'evna Bogdanovskaia," p 252.

100. Newman, L. M. "The "New Woman" (1890–1915)." In *Men's Ideas/ Women's Realities: Popular Science, 1870–1915;* Newman L. M., Ed.; Pergamon: New York, 1984; pp 298–329.

101. Solomon, *In the Company of Educated Women,* p 120.

102. Cookingham, M. E. "Combining Marriage, Motherhood and Jobs Before World War II: Women College Graduates, Classes of 1905–1935." *J. Family Hist.* **1984** (Summer), *9,* 349–364.

103. Cited in Gordon, L. D. *Gender and Higher Education in the Progressive Era;* Yale University Press: New Haven, CT, 1990; p 2.

104. Cited in Gordon, *Gender and Higher Education in the Progressive Era,* p 80.

Chapter 5

1. Rayner-Canham, M. F.; Rayner-Canham, G. W. "Women's Fields of Chemistry: 1900–1920." *J. Chem. Educ.* **1996,** *73,* 136–138.

2. Rossiter, M. W. "'Womens Work' in Science, 1880–1910." *Isis* **1980,** *71,* 381–398; also in Rossiter, M. W. *Women Scientists in America: Struggles and Strategies to 1940;* Johns Hopkins University Press: Baltimore, MD, 1982; pp 53–57.

3. Gornick, V. *Women in Science;* Simon & Schuster: New York, 1983; p 15.

4. Bonta, M. M. *Women in the Field: America's Pioneering Women Naturalists;* Texas A&M Press: College Station, TX, 1991; pp xii–xiii.

5. The importance of mentors in the encouragement of women in science is described in Didion, C. J. "Mentoring Women in Science." *Educ. Horizons* **1995** (Spring), 141–144; Gibbons, A. "Key Issue: Mentoring." *Science (Washington, D.C.)* **1992,** *255,* 1368; Yentsch, C. M.; Sindermann, C. J. *The Woman Scientist: Meeting the Challenges for a Successful Career;* Plenum: New York, 1992.

6. Ewald, P. P. "The Beginnings." In *Fifty Years of X-ray Diffraction;* Ewald, P. P., Ed.; International Union of Crystallography: Utrecht, Netherlands, 1962; pp 6–80; Gasman, L. D. "Myths and X-rays." *Br. J. Phil. Sci.* **1975,** *26,* 51–60; Speakman, J. C. "The Discovery of X-ray Diffraction." *J. Chem. Educ.* **1980,** *57,* 489–490.

7. Caroe, G. B. *William Henry Bragg 1862–1942: Man and Scientist;* Cambridge University Press: Cambridge, UK, 1978; Phillips, D. "William Lawrence Bragg." *Biog. Mem. Fellows R. Soc.* **1979,** *25,* 75–143.

8. Julian, M. M. "Crystallography at the Royal Institution." *Chem. Br.* **1986,** *22,* 729–732.

9. Julian, M. M. "Women in Crystallography." In *Women of Science: Righting the Record;* Kass-Simon, G.; Farnes, P., Eds.; Indiana University Press: Bloomington, IN, 1990; pp 335–383.

10. Julian, "Women in Crystallography," p 342.

11. Anne Sayre to Maureen Julian, cited in Julian, "Women in Crystallography," pp 339–340.

12. W. H. Bragg to Sir Richard Gregory, 3 July 1934, W. H. Bragg Archives, The Royal Institution.

13. Lomer, W. M. "Blowing Bubbles with Bragg." In *Selections and Reflections: The Legacy of Sir Lawrence Bragg;* Thomas, J. M.; Phillips, D., Eds.; Royal Institution of Great Britain: London, 1990; p 117.

14. Portugal, F. H.; Cohen, J. S. *A Century of DNA;* Massachusetts Institute of Technology Press: Cambridge, MA, 1977; p 267.

15. Warner, D. J. "Women Astronomers." *Natural Hist.* **1979,** 88, 12–26; Dobson, A.; Bracher, K. K. "A Historical Introduction to Women in Astronomy." *Mercury* **1992,** *21,* 4–15; Fraknoi, A.; Freitag, R. "Women in Astronomy: A Bibliography." *Mercury* **1992,** *21,* 46–47; Lankford, J.; Slavings, R. L. "Gender and Science: Women in American Astronomy, 1859–1940." *Phys. Today* **1990** (Mar), *43,* 58–65.

16. Rossiter, M. W. "'Womens Work' in Science, 1880–1910." *Isis* **1980,** *71,* 381–398; also in Rossiter, M. W. *Women Scientists in America;* Johns Hopkins University Press: Baltimore, MD, 1982; pp 53–57.

17. Fleming, W. "A Field for 'Women's Work' in Astronomy." *Astron. Astrophys.* **1893,** *12,* 688–689.

18. Lankford, J.; Slavings, R. L. "Gender and Science: Women in American Astronomy, 1859–1940." *Phys. Today* **1990** (Mar), *43,* 58–65.

19. Lonsdale, K. "Women in Science: Reminiscences and Reflections." *Impact Sci. Soc.* **1970,** *20,* 45–59.

20. Julian, M. M. "Profiles in Chemistry: Kathleen Lonsdale 1903–1971." *J. Chem. Educ.* **1982,** *59,* 965–966; Julian, M. M. "X-ray Crystallography and the Work of Dame Kathleen Lonsdale." *Phys. Teacher* **1981,** *3,* 159–165; Julian, M. M. "Kathleen and Thomas Lonsdale: Forty-Three Years of Spiritual and Scientific Life Together." In *Creative Couples in Science;* Pycior, H.; Slack, N. G.; Abir-Am, P. G., Eds.; Rutgers University Press: Brunswick, NJ, 1995.

21. Tuke, M. J. *A History of Bedford College for Women: 1849–1937;* Oxford University Press: Oxford, UK, 1939.

22. Lonsdale, "Women in Science: Reminiscences and Reflections."

23. Hodgkin, D. M. C. "Kathleen Lonsdale." *Biog. Mem. Fellows R. Soc.* **1976,** 446–484.

24. Cited in Hodgkin, "Kathleen Lonsdale," p 449.

25. Lonsdale, K. *Simplified Structure Factor and Electron Density Formulae for the 230 Space-Groups of Mathematical Crystallography;* G. Bell & Sons: London, 1936.

26. K. Lonsdale to W. H. Bragg, 16 November 1927, W. H. Bragg Archives, The Royal Institution.

27. Julian, M. M. "Kathleen Lonsdale and the Planarity of the Benzene Ring." *J. Chem. Educ.* **1981,** *58,* 365–366.

28. Vanderbilt, B. "Kekulé's Whirling Snake: Fact or Fiction." *J. Chem. Educ.* **1975,** *52,* 709; Wotiz, J. H.; Rudolfesky, S. "Kekulé's Dreams: Fact or Fiction?." *Chem. Br.* **1984,** *20,* 720–723.

29. Frondel, C.; Marvin, U. B. "Lonsdaleite, a Hexagonal Polymorph of Diamond." *Nature (London)* **1967,** *214,* 587–589.

30. *International Tables for X-ray Crystallography;* Lonsdale, K.; Henry, N. F. M., Eds.; Kynoch Press: Birmingham, UK, 1952; Vol. I; Lonsdale, K.; Kasper, J., Eds.; 1959; Vol. II; Lonsdale, K.; MacGillavry, C. H.; Reich, G. D., Eds.; Vol. III.

31. Mason, J. "The Admission of the First Women to the Royal Society of London." *Notes Records Roy. Soc. (London)* **1992,** *46,* 279–300; Mason, J. "Hertha Ayrton (1854–1923) and the Admission of Women to the Royal Society of London." *Notes Records R. Soc. (London)* **1991,** *45,* 201–220.

32. Lonsdale, K.; Milledge, H. J.; Nave, E.; Weller, F. H. "Transformation of Cubic Boron Nitride to a Graphitic Form of Hexagonal Boron Nitride." *Nature (London)* **1959,** *184,* 1545–1549.

33. Kathleen Lonsdale to A.V. Hill, 7 June 1953, A. V. Hill collection, Churchill College, Cambridge University.

34. Lonsdale, K. *Is Peace Possible?;* Penguin: London, 1957.

35. Cited in Hodgkin, "Kathleen Lonsdale," p 474.

36. Thomas Lonsdale to Sir Lawrence Bragg, 24 May 1971, Lawrence Bragg papers, Royal Institution. For a discussion of the relationship between Kathleen and Thomas Lonsdale, see Julian, M. M. "Kathleen and Thomas Lonsdale: Forty-Three Years of Spiritual and Scientific Life Together." In *Creative Couples in the Sciences;* Pycior, H. M., Slack, N. G.; Abir-Am, P. G., Eds.; Rutgers University Press: New Brunswick, NJ, 1996; pp 170–181.

37. *See,* for example, Lonsdale, K.; Sutor, D. J.; Wooley, S. "Composition of Urinary Calculi by X-ray Diffraction. Collected Data from Various Localities: Part I." *Br. J. Urol.* **1968,** *40,* 33–36; Lonsdale, K.; Sutor, D. J.; Wooley, S. "Composition of Urinary Calculi by X-ray Diffraction. Collected Data from Various Localities: Part II." *Br. J. Urol.* **1968,** *40,* 402–411; Lonsdale, K. "Human Stones," *Science (Washington, D.C.)* **1968,** *159,* 1199–1207.

38. Thomas Lonsdale to Mr. Glanville, undated, Lawrence Bragg papers, Royal Institution.

39. The biographical notes on Dorothy Crowfoot Hodgkin were mostly

derived from McGrayne, S. B. *Nobel Prize Women in Science;* Birch Lane Press: New York, 1993; pp 225–254; Julian, M. N. "Profiles in Chemistry: Dorothy Crowfoot Hodgkin, Nobel Laureate." *J. Chem. Educ.* **1982,** *59,* 124–125; Goldwhite, H. "Dorothy Mary Crowfoot Hodgkin (1910–)." In *Women in Chemistry and Physics: A Biobibliographic Sourcebook;* Grinstein, L. S.; Rose, R. K.; Rafailovich, M. H., Eds.; Greenwood Press: Westport, CT, 1993, pp 253–260; Farago, P. "Impact: Interview with Dorothy Crowfoot Hodgkin." *J. Chem. Educ.* **1977,** *54,* 214–216; Perutz, M. F. "Dorothy Hodgkin" address delivered at a memorial service in the University Church, Oxford on 4 March 1995 (Dr. Savitri Chandrasekhar is thanked for a copy of the address).

40. Hudson, G. "Unfathering the Thinkable: Gender, Science and Pacifism in the 1930s." In *Science and Sensibility: Gender and Scientific Enquiry, 1780–1945;* Benjamin, M., Ed.; Blackwell: Oxford, UK, 1991; p 275.

41. Julian, "Women in Crystallography," p 376.

42. Hodgkin, D. C. "J.D. Bernal." *Biog. Mem. Fellows R. Soc.* **1980,** 17–84.

43. Abir-Am, P. G. "Women in Research Schools: Approaching an Analytical Lacuna in the History of Chemistry and Allied Sciences." In *Chemical Sciences in the Modern World;* Mauskopf, S. H., Ed.; University of Pennsylvania Press: Philadelphia, PA, 1993; p 386.

44. Hodgkin, "J.D. Bernal," p 31.

45. Riley, D. P. "Oxford: the Early Years." In *Structural Studies on Molecules of Biological Interest, A Volume in Honour of Professor Dorothy Hodgkin;* Dodson, G.; Glusker, J.; Sayre, D., Eds.; Clarendon Press: Oxford, UK, 1981; p 17. Reprinted by permission of Oxford University Press.

46. Riley, D. P. "Oxford: the Early Years," p 17.

47. Perutz, M. "Forty Years' Friendship with Dorothy." In *Structural Studies on Molecules of Biological Interest, A Volume in Honour of Professor Dorothy Hodgkin;* Dodson, G.; Glusker, J.; Sayre, D., Eds.; Clarendon Press: Oxford, UK, 1981; p 6. Reprinted by permission of Oxford University Press.

48. Carlisle, C. H.; Crowfoot, D. "The Crystal Structure of Cholestryl Iodide." *Proc. R. Soc.* **1945,** *A184,* 64–83.

49. Ramaseshan, S. "Dorothy Hodgkin and the Indian Connection" *Notes Records R. Soc. London* 1996, *50,* 115–127.

50. Cited in McGrayne, *Nobel Prize Women in Science;* p 244.

51. Trueblood, K. N. "Structure Analysis by Post and Cable." In *Structural Studies on Molecules of Biological Interest, A Volume in Honour of Professor Dorothy Hodgkin;* Dodson, G.; Glusker, J.; Sayre, D., Eds.; Clarendon Press: Oxford, UK, 1981; p 87.

52. Perutz, M. "Forty Years' Friendship with Dorothy." In *Structural Studies on Molecules of Biological Interest, A Volume in Honour of Professor Dorothy Hodgkin;* Dodson, G.; Glusker, J.; Sayre, D., Eds.; Clarendon Press: Oxford, UK, 1981; p 10.

53. Robertson, J. H. "Memories of Dorothy Hodgkin and of the B12 Structure in 1951–4." In *Structural Studies on Molecules of Biological Interest, A Volume in Honour of Professor Dorothy Hodgkin;* Dodson, G.; Glusker, J.; Sayre,

D., Eds.; Clarendon Press: Oxford, UK, 1981, p 73. Reprinted by permission of Oxford University Press.

54. Hudson, G. "Unfathering the Thinkable: Gender, Science and Pacifism in the 1930s." In *Science and Sensibility: Gender and Scientific Enquiry, 1780–1945;* Benjamin, M., Ed.; Blackwell: Oxford, UK, 1991; p 281.

55. Dunitz, J. D. "Organic Chemistry, X-ray Analysis and Dorothy Hodgkin." In *Structural Studies on Molecules of Biological Interest, A Volume in Honour of Professor Dorothy Hodgkin;* Dodson, G.; Glusker, J.; Sayre, D., Eds.; Clarendon Press: Oxford, UK, 1981; p 59. Reprinted by permission Oxford University Press.

56. The biographical notes on Rosalind Franklin were mostly derived from Sayre, A. *Rosalind Franklin and DNA;* W. W. Norton: New York, 1975; McGrayne, S. B. *Nobel Prize Women In Science;* Birch Lane Press: New York, 1993, pp 304–332; Glynn, J. "Rosalind Franklin 1920–1958." In *Cambridge Women: Twelve Portraits;* Shils, E.; Blacker, C., Eds.; Cambridge University Press: Cambridge, UK, 1996; pp 267–282; Hubbard, R. "Reflections on the Story of the Double Helix." *Women's Stud. Int. Quart.* **1979**, *2*, 261–273; Julian, M. M. "Profiles in Chemistry: Rosalind Franklin, From Coal to DNA to Plant Viruses." *J. Chem. Educ.* **1983**, *60*, 660–662; Strasser, J. "Jungle Law: Stealing the Double Helix." *Sci. People* **1976** (Sept/Oct), *8*, 29–31. *See also* Richards, E.; Schuster, J. "The Feminine Method as Myth and Accounting Resource: A Challenge to Gender Studies and Social Studies of Science." *Soc. Stud. Sci.* **1989**, *19*, 697–720.

57. Franklin, M. *Rosalind;* n.p.: Frome, UK, undated, p 5.

58. McGrayne, *Nobel Prize Women in Science;* p 308.

59. Sayre, *Rosalind Franklin and DNA;* p 58.

60. Cited in Sayre, *Rosalind Franklin and DNA;* p 58.

61. Julian, "Women in Crystallography," p 345.

62. M. Franklin, *Rosalind;* p 12.

63. Cited in McGrayne, *Nobel Prize Women in Science;* p 312.

64. Cited in McGrayne, *Nobel Prize Women in Science;* p 313.

65. Judson, H. F. "Annals of Science: The Legend of Rosalind Franklin." *Sci. Digest* **1986**, *94*, 56–59, 78–83.

66. Judson, "Annals of Science: The Legend of Rosalind Franklin," p 79.

67. Watson, J. D.; Crick, F. H. C. "A Structure for Deoxyribose Nucleic Acid." *Nature (London)* **1953**, *171*, 737–738. For a more complete account of the DNA saga, *see* Olby, R. *The Path to the Double Helix;* Macmillan: London, 1974.

68. Cited in Julian, M. M. "Women in Crystallography," p 363.

69. K. Lonsdale to R. Franklin, 4 January 1955, Franklin Archives, Churchill College, Cambridge University (CC–CU).

70. Mackay, A. L. "The Lab." *Chem. Intell.* **1995**, *1(1)*, 12–18.

71. Watson, J. D. "The Double Helix." In *The Double Helix: A New Critical Edition;* Stent, G. S. Ed.; Weidenfeld and Nicolson: London, 1968. This version includes additional commentaries, several of which contradict Watson's viewpoint.

72. Sayre, *Rosalind Franklin and DNA;* p 18. Copyright © 1975 by Anne Sayre.

Reprinted by permission of W. W. Norton & Company, Inc.

73. *See* Klug, A. "Rosalind Franklin and the Discovery of the Structure of DNA." *Nature (London)* **1968**, *219*, 808–810, 843–844; Klug, A. "Rosalind Franklin and the Double Helix." *Nature (London)* **1974**, *248*, 787.

74. A. J. Caraffi to A. Klug, 29 August 1968, Franklin Archives, CC–CU.

75. Bernal, J. D. "Dr. Rosalind E. Franklin." *Nature (London)* **1958**, *182*, 154. Reprinted with permission, © 1958 Macmillan Magazines Limited.

76. A. Klug to P. Siekevitz (New York Academy of Sciences), 14 April 1976, Franklin Archives, Churchill College, Cambridge University.

77. Rose, H. "Nine Decades, Nine Women, Ten Nobel Prizes: Gender at the Apex of Science." In *Love, Power and Knowledge: Towards a Feminist Transformation of the Sciences;* Rose, H.. Ed.; Polity Press: Cambridge, UK, 1994; p 157.

78. Julian, M. M. "Profiles in Chemistry: Isabella Karle and a Mathematical Breakthrough in Crystallography." *J. Chem. Educ.* **1986**, *63*, 66–67; Roscher, N. M. "Isabella Helen Lugoski Karle (1921–)." In *Women in Chemistry and Physics: A Biobibliographic Sourcebook;* Grinstein, L. S.; Rose, R. K.; Rafailovich, M. H., Eds.; 1993; pp 284–298. Karle was named by the *Science Citation Index* as among the 1000 most-cited world scientists, and, apart from her own awards, she contributed to the work for which her husband was to receive the Nobel Prize in 1985.

79. Julian, M. M. "Women in Crystallography," p 336.

80. Cited in O'Driscoll, C. "Minorities and Mentors: The X-ray Visionaries." *Chem. Br.* **1996** (Feb), *32*, 5–8.

Chapter 6

1. Nye, M. J. "N-rays: an Episode in the History and Psychology of Science." *Hist. Stud. Phys. Sci.* **1980**, *11*, 125–146.

2. Badash, L. "Radioactivity before the Curies." *Am. J. Phys.* **1965**, *33*, 128–135; Badash, L. "The Discovery of Radioactivity." *Phys. Today* **1996**, *49(2)*, 21–26.

3. Badash, L. "The Discovery of Thorium's Radioactivity." *J. Chem. Educ.* **1966**, *43*, 219–220.

4. *Radioactivity and Atomic Theory (Annual Progress Reports on Radioactivity 1904–1920 to the Chemical Society by Frederick Soddy F.R.S.);* Trenn, T. J., Ed.; Taylor & Francis: London, 1975.

5. Lucas, R. *Bibliographie der radioactiven Stoffe;* Verlag von Leopold Voss: Hamburg and Leipzig, Germany, 1908.

6. Lawson, R. W. "The Part Played by Different Countries in the Development of the Science of Radioactivity." *Scientia* **1921**, *30*, 257–270; Jauncey, G. E. M. "The Early Years of Radioactivity." *Am. J. Phys.* **1946**, *14*, 226–241. The three schools of radioactivity research are discussed more fully by Rayner-Canham, M. F.; Rayner-Canham, G. W. "Early Years of Radioactivity." In *A Devotion to Their Science: Pioneer Women of Radioactivity;* Rayner-Canham, M. F.; Rayner-Canham, G. W., Eds.; Chemical Heritage Foundation, Philadelphia, PA, and McGill–Queen's University Press: Montreal,

PQ, 1997; pp 4–6. The differences between some chemical schools are given by Nye, M. J. "National Styles? French and English Chemistry in the Nineteenth and Early Twentieth Centuries." *Osiris* **1993**, *8*, 30–49.

7. Malley, M. "The Discovery of Atomic Transmutation: Scientific Styles and Philosophies in France and Britain." *Isis* **1979**, *70*, 213–223.

8. For a complete coverage of all thirty women who were published researchers in the early studies of radioactivity, *see* Rayner-Canham, M. F.; Rayner-Canham, G. W., Eds.; *A Devotion to Their Science: Pioneer Women of Radioactivity.*

9. Price, D. de S. *Little Science, Big Science . . . And Beyond;* Columbia University Press: New York, 1986; pp 119–134; Crane, D. *Invisible Colleges;* University of Chicago Press: Chicago, IL, 1972.

10. Gleditsch, E. to Meitner, L. 16 June 1926, Meitner Collection, Churchill College, Cambridge University.

11. Rayner-Canham, M. F.; Rayner-Canham, G. W. "Elizabeth Róna: The Polonium Woman." In *A Devotion to Their Science: Pioneer Women of Radioactivity;* Rayner-Canham M. F.; Rayner-Canham, G. W., Eds.; pp 209–216.

12. Ramstedt, E. to Curie, M., 11 August 1925, Bibliothèque Nationale.

13. The original quote was by George Herbert Palmer in *The Life of Alice Freeman Palmer;* Houghton Mifflin: Boston, MA, 1924; pp 172–173. Cited in Barnard, J. *Academic Women;* Pennsylvania State University Press: University Park, PA, 1964; p 207. The "marriage question" is also discussed by Glazer, P.; Slater, M. *Unequal Colleagues: The Entrance of Women into the Professions, 1890–1940;* Rutgers University Press: Brunswick, NJ, 1986; pp 57–64.

14. Kubanek, A-M. W.; Grzegorek, G. P. "Ellen Gleditsch: Professor and Humanist." In *A Devotion to Their Science: Pioneer Women of Radioactivity;* Rayner-Canham, M. F.; Rayner-Canham, G. W., Eds.; pp 51–75.

15. Leslie, M. S. to Smithalls, A., 30 November 1909, Smithalls Collection, University of Leeds. *See also* Rayner-Canham, G. W.; Rayner-Canham, M. F. "A Chemist of Some Repute." *Chem. Br.* **1993**, *29*, 206–208.

16. Pflaum, R. *Grand Obsession: Madame Curie and Her World;* Doubleday: New York, 1991; p 82.

17. Rayner-Canham, M. F.; Rayner-Canham, G. W. *Harriet Brooks: Pioneer Nuclear Scientist;* McGill-Queen's University Press: Montreal, PQ, 1992.

18. Golonka, M. C.; Róziewicz, J.; Starosta, J.; Tokhadze, K. G. "Jadwiga Szmidt (1889–1940): A Pioneer Woman in Nuclear and Electrotechnical Sciences." *Am. J. Phys.* **1994**, *62*, 947–948; Rayner-Canham, M. F.; Rayner-Canham, G. W. "Jadwiga Szmidt: A Passion for Science." In *A Devotion to Their Science: Pioneer Women of Radioactivity;* Rayner-Canham M. F.; Rayner-Canham, G. W., Eds.; pp 145–151.

19. Heisch, A. "Queen Elizabeth I and the Persistence of Patriarchy." *Feminist Rev.,* **1980**, *4*, 45–55.

20. Crowther, J. G. *The Cavendish Laboratory 1874–1974;* Science History Publications: New York, 1974; p. 91.

21. Ernest Rutherford and William Pope, to The Editor, *The Times* (London); 8 December 1920, p 8.

22. Rayner-Canham, M. F.; Rayner-Canham, G. W. "Women's Fields of Chemistry: 1900–1920," *J. Chem. Educ.* **1996**, *73*, 136–138.

23. Kapitza, P. L. "Recollections of Lord Rutherford." *Proc. R. Soc. (London)* **1966**, *A294*, 123–137.

24. Frisch, O. R. "How It All Began." In *History of Physics: Readings from Physics Today 2;* Weart, S. R.; Phillips, M., Eds.; American Institute of Physics: New York, 1985; p 272. Abstracted with permission ©1985 American Institute of Physics.

25. Halpern, L. "Marietta Blau: Discoverer of Cosmic Ray 'Stars'." In *A Devotion to Their Science: Pioneer Women of Radioactivity;* Rayner-Canham, M. F.; Rayner-Canham, G. W., Eds.; McGill–Queen's University Press: Montreal, PQ, 1997; pp 196–204.

26. Wilson, D. *Rutherford: Simple Genius;* Massachusetts Institute of Technology Press: Cambridge, MA, 1983; p 46.

27. Paneth, F. A. "Prof. Stefan Meyer," *Nature (London)* **1950**, *165*, 548–549. Reprinted with permission © 1950 Macmillan Magazines Limited.

28. Lawson, R. W. "Prof. Stefan Meyer," *Nature (London)* **1950**, *165*, 549. Lawson had been trapped in Vienna at the outbreak of the First World War. Meyer rescued Lawson from internment, kept him as a "house guest" for the duration of the war, gave him financial support, and provided Lawson with research facilities at the University. Reprinted with permission © 1950 Macmillan Magazines Limited.

29. Róna, E. *How It Came About;* Oak Ridge Associated Universities: Oak Ridge, TN, 1978; p 15.

30. Curie, E. *Madame Curie;* Doubleday: New York, 1937; Reid, R. *Marie Curie,* Saturday Review: New York, 1974; Giroud, F. *Marie Curie—A Life;* Holmes and Meier: New York, 1986; Pflaum, R. *Grand Obsession: Marie Curie and Her World;* Doubleday: New York, 1989; Quinn, S. *Marie Curie: A Life;* Simon & Schuster: New York, 1995. Unless otherwise referenced, this account was assembled from these sources.

31. Badash, L. "Book Review: Decay of a Radioactive Halo." *Isis* **1975**, *66*, 566–568.

32. Brush, S. G. "Women in Physical Science: From Drudges to Discoverers." *Phys. Teach.* **1985**, *23*, 11–19.

33. Pycior, H. M. "Reaping the Benefits of Collaboration While Avoiding Its Pitfalls: Marie Curie's Rise to Scientific Prominence." *Soc. Stud. Sci.* **1993**, *23*, 301–323. Pycior has also argued that Pierre Curie had for many years, before he met Marie, collaborated with his brother, Jacques. To work with Marie, therefore, was a continuation of his collaborative style. *See* Pycior, H. M. "Pierre Curie and 'His Eminent Collaborator Mme Curie': Complementary Partners." In *Creative Couples in the Sciences;* Pycior, H. M., Slack, N. G.; Abir-Am, P. G., Eds.; Rutgers University Press: New Brunswick, NJ, 1996; pp 39–56.

34. Curie, Mme. S. "Propriétés magnétiques des aciers trempés." *Comptes Rendus* **1897**, *125*, 1165–1169.

35. Wolke, R. L. "Marie Curie's Doctoral Thesis: Prelude to a Nobel Prize." *J. Chem. Educ.* **1988**, *65*, 561–573.

36. Badash, L. "The Discovery of Thorium's Radioactivity." *J. Chem. Educ.* **1966,** *43,* 219–220.

37. Curie, Mme "Rayons émis par les composés de l'uranium et du thorium." *Comptes Rendus* **1898,** *126,* 1101–1103.

38. Curie, P.; Curie, S. "Sur une substance nouvelle radioactive, contenue dans la pechblende." *Comptes Rendus* **1898,** *127,* 175–178.

39. Curie, P.; Curie, S.; Bémont, M. G. "Sur une nouvelle substance fortement radioactive, contenue dans la pechblende." *Comptes Rendus* **1898,** *127,* 1215–1217.

40. Cited in Weeks, M. E. "The Discovery of the Elements. XIX. The Radioactive Elements." *J. Chem. Educ.* **1933,** *10,* 79–90.

41. Curie, Mme. "Sur le poids atomique du radium." *Comptes Rendus* **1902,** *135,* 161–163.

42. Cited by Weeks, "The Discovery of the Elements. XIX," p 81.

43. Pycior, H. M. "Reaping the Benefits of Collaboration While Avoiding Its Pitfalls."

44. Cited by Pycior, H. M. "Reaping the Benefits of Collaboration While Avoiding Its Pitfalls." p 308.

45. Debierne, A. "Sur une nouvelle matière radio-active." *Comptes Rendus* **1899,** *129,* 593–595; Debierne, A. "Sur un nouvel élément radioactif: actinium." *Comptes Rendus* **1900,** *130,* 906–908.

46. Koblitz, A. H. *A Convergence of Lives: Sofia Kovalevskaia, Scientist, Writer, Revolutionary;* Birkhäuser: Boston, MA, 1983.

47. Quinn comments "By mid-July of 1910, all the evidence suggests, Marie and Paul had become lovers." *See* Quinn, *Marie Curie: A Life,* p 262. Pflaum is more circumspect in her biography, but she notes Langevin's reputation as being "dangerous for young girls." Langevin later fathered a child by one of Marie Curie's research students, Elie Montel. *See* Pflaum, *Grand Obsession: Marie Curie and Her World,* p 166.

48. Couture-Cherki cites a 1972 quote comparing Pierre and Marie Curie: "Between Pierre and Marie Curie, Pierre Curie was a creator whose very genius established new laws of physics. Marie Curie radiated other qualities: her character, her exceptional tenacity, her precision, and her patience." *See* Couture-Cherki, M. "Women in Physics." In Rose, H.; Rose, S. *The Radicalisation of Science: Ideology of/in the Natural Sciences;* Macmillan, London, 1976; p 73.

49. Pflaum, *Grand Obsession: Madame Curie and Her World,* p 160.

50. Quinn, *Marie Curie: A Life,* p 343.

51. McGrayne, S. B. *Nobel Prize Women in Science;* Birch Lane Press: New York, 1993; p 34.

52. Reed, *Marie Curie,* pp 276–279. Elizabeth Róna has described the careless handling of radioactive materials in the Curie laboratories. *See* Róna, E. "Laboratory Contamination in the Early Period of Radioactivity Research." *Health Phys.* **1979,** *37,* 723–727.

53. Curie, M. *Pierre Curie,* Macmillan: New York, 1923; p 197.

54. Dupuy, G. "Notice sur la vie et les travaux de André Debierne (1874–1949)." *Bull. Soc. Chim. Fr.* **1950,** 1024–1026.

55. Kronen, T.; Pappas, A. C. *Ellen Gleditsch. Et liv i forskning og medmenneske-lighet;* Aventura Forlag: Oslo, Norway, 1987; p 133.

56. Kubanek, A-M. W.; Grzegorek, G. P. "Ellen Gleditsch: Professor and Humanist." In *A Devotion to Their Science: Pioneer Women of Radioactivity;* Rayner-Canham, M. F.; Rayner-Canham, G. W., Eds.; pp 51–75.

57. Gleditsch, E. "Sur quelques derivés d'amylbenzene tertiaire." *Bull. Soc. Chim. Fr.* **1907,** *35,* 1094–1097 .

58. Marie Curie to Ellen Gleditsch, 26 July 1907, Oslo University Library (OUL).

59. Curie, M.; Gleditsch, E. "Action de l'émanation du radium sur les solutions des sels de cuivre." *Comptes Rendus* **1908,** *147,* 345–349 .

60. Gleditsch, E. "Sur le rapport entre l'uranium et le radium dans les minéraux radioactifs." *Le Radium* **1911,** *8,* 256–273 .

61. Bertram Boltwood to Ernest Rutherford, 12 September 1913; in *Rutherford and Boltwood: Letters on Radioactivity;* Badash, L. Ed.; Yale University Press: New Haven, CT, 1969; p 285.

62. Badash, L. *Radioactivity in America: Growth and Decay of a Science;* Johns Hopkins University Press: Baltimore, MD, 1979; pp 153–158.

63. Bertram Boltwood to Ellen Gleditsch, 2 July 1914, OUL.

64. Gleditsch, E. "The Life of Radium." *Am. J. Sci.* **1916,** *41,* 112–124 .

65. Forbes, G. S. "Investigations of Atomic Weights by Theodore William Richards." *J. Chem. Educ.* **1932,** *9,* 453–458; H. S. King, "Pioneering Research on Isotopes at Harvard." *J. Chem. Educ.* **1959,** *36,* 225–227.

66. Soddy, F. "The Origins of the Conceptions of Isotopes." Nobel Lecture 12 December 1922; In *Nobel Lectures—Chemistry 1901–1921;* Elsevier: New York, 1964; pp 371–399.

67. Richards, T. W.; Wadsworth, C. "Further Study of the Atomic Weight of Lead of Radioactive Origin." *J. Am. Chem. Soc.* **1916,** *38,* 2613–2622.

68. Anders, O. U. "The Place of Isotopes in the Periodic Table: The 50th Anniversary of the Fajans–Soddy Displacement Laws." *J. Chem. Educ.* **1964,** *41,* 522–525.

69. Marie Curie to Ellen Gleditsch, 22 June 1916, OUL.

70. Gleditsch, E.; Ramstedt, E. *Radium og de radioaktive processer;* Aschehoug Forlag: Kristiania, Norway, 1917.

71. Rayner-Canham, M. F.; Rayner-Canham, G. W. ". . . And Yet More French Women!." In *A Devotion to Their Science: Pioneer Women of Radioactivity;* Rayner-Canham, M. F.; Rayner-Canham, G. W., Eds.; pp 124–126.

72. Gleditsch, E. *Contribution to the Study of Isotopes;* J. Dybwad: Oslo, Norway, 1925.

73. Rayner-Canham, M. F.; Rayner-Canham, G. W. "Catherine Chamié: Devoted Researcher of the Institut de Radium." In *A Devotion to Their Science: Pioneer Women of Radioactivity;* Rayner-Canham, M. F.; Rayner-Canham, G. W., Eds.; pp 82–86.

74. Cited in Quinn, *Marie Curie: A Life;* p 420.

75. Róna, E. *How It Came About;* Oak Ridge Associated Universities: Oak Ridge, TN, 1978; p 28.

76. McKowan, R. *She Lived for Science. Irène Joliot-Curie;* Julian Messner: New York, 1992; Crossfield, T. "Irène Joliot-Curie: Following in Her Mother's

Footsteps." In *A Devotion to Their Science: Pioneer Women of Radioactivity;* Rayner-Canham, M. F.; Rayner-Canham, G. W., Eds.

77. In Ziegler, G., Ed., *Choix de Lettres de Marie Curie et Irène Joliot-Curie;* Les Éditeurs Français Réunis, Paris, 1974; p 84.

78. Curie, I. "Sur les poids atomique du chlore dans quelques minéraux." *Comptes Rendus* **1921,** *172,* 1025–1028.

79. Curie, I. "Recherches sur les rayons α du polonium. Oscillations de parcours, vitesse d'émission, pouvior ionisant." *Ann. Phys.* **1925,** *2,* 403.

80. Cited in Goldsmith, M. *Frédérick Joliot-Curie. A Biography;* Lawrence and Wishart: London, 1976; p 31. Bensaude-Vincent notes that "Both Irène and [her daughter] Hélène married within the 'tribe' of the scientific elite [Hélène married the grandson of Paul Langevin], thus consolidating the French tradition of 'scientific endogamy,' since intermarriage between scientists' families was already frequent in early-nineteenth-century Europe." *See* Bensaude-Vincent, B. "Star Scientists in a Nobelist Family: Irène and Frédéric Joliot-Curie." In *Creative Couples in the Sciences;* Pycior, H. M.; Slack, N. G.; Abir-Am, P. G., Eds.; Rutgers University Press: New Brunswick, NJ, 1996; pp 57–71.

81. Chadwick, J. "Possible Existence of a Neutron." *Nature (London)* **1932,** *129,* 312; Oliphant, M. "The Beginning: Chadwick and the Neutron." *Bull. At. Sci.* **1982,** *38,* 14–18.

82. Curie, I.; Joliot, F. "Artificial Production of a New Kind of Radioelement." *Nature (London)* **1934,** *133,* 201–202.

83. Curie, I.; Savitch, P. "Sur le radioélément de périod 3,5 h. formé dans l'uranium par les neutrons." *Comptes Rendus* **1938,** *206,* 1643–1644.

84. Rossiter, M. W. " 'But She's an Avowed Communist!' L'Affaire Curie at the American Chemical Society, 1953–1955." *Bull. Hist. Chem.* **1997,** *(20),* 33–41.

85. Kauffman, G. B.; Adloff, J. P. "Marguerite Perey and the Discovery of Francium." *Educ. Chem.* **1989,** *26,* 135–137.

86. Perey, M. "Sur un élément 87, dérivé de l'Actinium." *Comptes Rendus* **1939,** *208,* 97–99.

87. Perey, M. "Francium." In *Nouveau traité de chimie minérale;* Pascal, P., Ed.; Masson: Paris, 1957; Vol. 3, pp 131–141.

88. Rayner-Canham, M. F.; Rayner-Canham, G. W. "Fanny Cook Gates, Physicist (1872–1931): A Lesson from the Past." *J. Women Minorities Sci. Eng.* **1997,** *3(1–2),* 53–64; Rayner-Canham, M. F.; Rayner-Canham, G. W. "Fanny Cook Gates: A Promise Unfulfilled." In *A Devotion to Their Science: Pioneer Women of Radioactivity;* Rayner-Canham, M. F.; Rayner-Canham, G. W., Eds.; pp 138–144.

89. Golonka, Róziewicz, Starosta and Tokhadze. "Jadwiga Szmidt (1889–1940): A Pioneer Woman in Nuclear and Electrotechnical Sciences." In *A Devotion to Their Science: Pioneer Women of Radioactivity;* Rayner-Canham, M. F.; Rayner-Canham, G. W., Eds., pp 145–151; Rayner-Canham, M. F., Rayner-Canham, G. W. "Jadwiga Szmidt: A Passion for Science".

90. Rayner-Canham, M. F.; Rayner-Canham, G. W. *Harriet Brooks: Pioneer Nuclear Scientist.*

91. Rutherford, E.; Brooks, H. T. "The New Gas from Radium." *R. Soc. Can. Trans.* **1901,** *3,* 21–25.

92. Harriet Brooks to Ernest Rutherford, undated, Cambridge University Archives (CUA).

93. Eve, A. S. "Some Scientific Centres. VIII The MacDonald Physics Building, McGill University, Montreal." *Nature (London)* **1906**, *74*, 272–275.

94. Cited in Flam, F. "Still a 'Chilly Climate' for Women?" *Science (Washington, D.C.)* **1991**, *252*, 1604–1606.

95. Brooks, H. "A Volatile Product from Radium." *Nature (London)* **1904**, *70*, 270.

96. Rutherford, E. "The Succession of Changes in Radioactive Bodies." *Phil. Trans. R. Soc.* **1904**, *A204*, 169–219.

97. Harriet Brooks to Laura Gill, 18 July 1906, Barnard College Archives (BCA).

98. Margaret Maltby to Laura Gill, 24 July 1906, BCA.

99. Laura Gill to Margaret Maltby, 30 July 1906, BCA.

100. Ernest Rutherford to Arthur Eve, May 1933. We thank Montague Cohen, McGill University, for a copy of this letter.

101. Halpern, L. "Marietta Blau: Discoverer of Cosmic Ray 'Stars.'"

102. Rayner-Canham, M. F.; Rayner-Canham, G. W. "Stefanie Horovitz: A Crucial Role in the Discovery of Isotopes." In *A Devotion to Their Science: Pioneer Women of Radioactivity;* Rayner-Canham, M. F.; Rayner-Canham, G. W., Eds.; pp 192–195.

103. Otto Hönigschmid to Lise Meitner, 27 June 1914, Meitner Archives, Churchill College Library, Cambridge University.

104. Leicester, H. M., Ed.; *Source Book in Chemistry 1900–1950;* Harvard University Press: Cambridge, MA, 1968; pp 82–84.

105. Kasimir Fajans to Elizabeth Róna, 31 August 1963, Bentley Historical Library, University of Michigan.

106. Sime, R. L. *Lise Meitner. A Life in Physics;* University of California Press: Berkeley, CA, 1996; Krafft, F. "Lise Meitner: Her Life and Times—On the Centenary of the Great Scientist's Birth." *Angew. Chem. (Int. Ed.)* **1978**, *17*, 826–842.

107. Watkins, S. A. "The Making of a Physicist." *Phys. Teach.* **1984**, *22*, 12–15.

108. Olga Steindler was the first woman to have obtained a doctorate in physics from the University of Vienna. *See* Sime, R. L. *Lise Meitner. A Life in Physics;* p 398, note 82.

109. Sime, R. L. "The Discovery of Protactinium." *J. Chem. Educ.* **1986**, *63*, 653–657.

110. Hahn, O.; Meitner, L. "Die Muttersubstanz des Actiniums, ein neues radioaktives Element von langer Lebensdauer." *Phys. Z.* **1918**, *19*, 208–218.

111. Meitner, L. "Über das Protoactinium." *Naturwissenschaften* **1918**, *6*, 324–326.

112. Sime, R. L. "Lise Meitner's Escape from Germany." *Am. J. Phys.* **1990**, *58*, 262–267. Abstracted with permission © 1990 American Association of Physics Teachers.

113. Sime, R. L. "Lise Meitner in Sweden 1938–1960: Exile from Physics." *Am. J. Phys.* **1994**, *62*, 695–701.

114. Otto Hahn to Lise Meitner, 19 December 1938, cited in Sime, R. L. "Lise Meitner and the Discovery of Fission." *J. Chem. Educ.* **1989**, *66*, 373–376.

115. Hahn. O.; Strassmann, F. "Über den Nachweis und das Verhalten der bei der Bestrahlung des Urans mittels Neutronen entstehenden Erdalkalimetalle." *Naturwissenschaften* **1939**, *27*, 11–15.

116. Stuewer, R. H. "The Origin of the Liquid-Drop Model and the Interpretation of Nuclear Fission." *Perspect. Sci.* **1994**, *2*, 76–129.

117. Frisch, O. R. "How It All Began," In Weart, S. R.; Phillips, M. *History of Physics: Readings from Physics Today 2;* p 276. *See also* Herrmann, G. "Five Decades Ago: From the 'Transuranics' to Nuclear Fission." *Angew. Chem. (Int. Ed.)* **1990**, *29*, 481–508.

118. Meitner, L.; Frisch, O. R. "Disintegration of Uranium by Neutrons: A New Type of Nuclear Reaction." *Nature (London)* **1939**, *143*, 239–240.

119. Hahn, O.; Strassmann, F. "Nachweis der Entstehung aktiver Bariumisotope aus Uran und Thorium durch Neutronenbestrahlung; Nachweis weiterer aktiver Bruchstücke bei der Uranspaltung." *Naturwissenschaften* **1939**, *27*, 89–95.

120. Sime, "Lise Meitner in Sweden," p 699. Sime provides a detailed discussion of Hahn's claim for discovery of nuclear fission in Sime, *Lise Meitner: A Life in Physics.*

121. Sime, R. "A Split Decision?" *Chem. Br.* **1994**, *30*, 482–484.

122. Hans Pettersson to Lise Meitner, 16 November 1945, Churchill College, Cambridge University; cited in Sime, "Lise Meitner in Sweden," p 698.

123. Dickman, S. "Meitner Receives Her Due." *Nature (London)* **1989**, *340*, 497.

124. Dagani, R. "ACS Adopts 'Seaborgium,' Other Names." *Chem. Eng. News* **1995**, *19 June*, 8.

125. For example, there is no reference to Noddack In *Women in Chemistry and Physics: A Biobibliographic Sourcebook;* Grinstein, L. S.; Rose, R. K., Rafailovich, M. H.; Eds.; Greenwood Press: Westport, CT, 1993.

126. Habashi, F. "Ida Noddack: Proposer of Nuclear Fission." In *A Devotion to Their Science: Pioneer Women of Radioactivity;* Rayner-Canham, M.F.; Rayner-Canham, G. W., Eds.; pp 217–225.

127. "Two New Elements of the Manganese Group." *Nature (London)* **1925**, *116*, 54–55.

128. Kenna, B. T. "The Search for Technetium in Nature." *J. Chem. Educ.* **1962**, *39*, 436–442; Rayner-Canham, G. W.; Pike, G. "The Search for the Elusive Element 43." *Educ. Chem.* **1993**, *30*, 12–14.

129. Noddack, I. "Über das Element 93." *Angew. Chem.* **1934**, *47*, 653–656.

130. Cited in Jungk, R. *Brighter Than a Thousand Suns;* Harcourt, Brace, Jovanovich: New York, 1958; p 62.

131. Sources on the life of Goeppert-Mayer include Dash, J. A. *A Life of One's Own;* Harper and Row: New York, 1973; Opfell, O. S. *The Lady Laureates;* Scarecrow: Metuchen, NJ, 1978; Sachs, R. G. "Maria Goeppert-Mayer—Two-Fold Pioneer." *Phys. Today* **1982** (Feb), *35*, 46–51; Johnson, K. E. "Maria Goeppert Mayer: Atoms, Molecules, and Nuclear Shells." *Phys. Today* **1986** (Sept), *39*, 44–49; Sachs, R. G. "Maria Goeppert Mayer, June 28, 1906–February 20, 1972." *Biog. Mem. Natl. Acad. Sci.* **1979**, *50*, 311–328.

132. Dash, *A Life of One's Own;* p 277.

133. Mayer. J. E.; Mayer, M. G. *Statistical Mechanics;* Wiley: New York, 1940.

134. McGrayne, *Nobel Prize Women in Science;* p 189.

135. Goeppert-Mayer, M. "The Shell Model." *Science (Washington, D.C.)* **1964,** *145,* 999–1006.

136. Rayner-Canham, G. W.; Rayner-Canham, M. F. "The Shell Model of the Nucleus." *Sci. Teach.* **1987,** *54(1),* 18–20; Baranger, E. U. "The Present Status of the Nuclear Shell Model." *Phys. Today* **1973,** *26,* 34–43; Seaborg, G. T. "Prospects for Further Considerable Extension of the Periodic Table." *J. Chem. Educ.* **1969,** *46,* 626–634.

137. Goeppert-Mayer, M.; Jensen, J. H. D. *Elementary Theory of Nuclear Shell Structure;* John Wiley: New York, 1955.

138. In the post-Second World War era, Moïse Haïssinsky at the Laboratoire Curie and Professor N. A. Bach of the Laboratory of Radiation Chemistry, Moscow State University, continued to employ significant numbers of women researchers. *See* Kroh, J., Ed.; *Early Developments in Radiation Chemistry;* Royal Society of Chemistry: London, 1989; pp 103, 362.

139. Cited in Badash, L. "Nuclear Physics in Rutherford's Laboratory Before the Discovery of the Neutron." *Am. J. Phys.* **1983,** *51,* 884–889. Abstracted with permission © 1983 American Association of Physics Teachers.

140. Badash, L. "The Suicidal Success of Radiochemistry." *Br. J. Hist. Sci.* **1979,** *12,* 245–256.

141. Geiger, R. L. *To Advance Knowledge: The Growth of American Research Universities, 1900–1940;* Oxford University Press: New York, 1986; pp 233–240; Bensaude-Vincent, B.; Stengers, I. *A History of Chemistry;* Harvard University Press: Cambridge, MA, 1996; pp 230–231.

142. Slater, J. C. "Quantum Physics Between the Wars." *Phys. Today* **1968,** 43–51. Abstracted with permission © 1968 American Institute of Physics.

143. Kevles, D. J. *The Physicists: The History of a Scientific Community;* Harvard University Press: Cambridge, MA, 1987; p 207.

144. Herzenberg, C. L.; Howes, R. H. "Women of the Manhattan Project" *Tech. Rev.* **1993,** *36*(Nov/Dec), 32–40.

145. Ajzenberg-Selove, F. *A Matter of Choices: Memoirs of a Female Physicist;* Rutgers University Press: New Brunswick, NJ, 1994; p 114.

146. Traweek, S. "High Energy Physics: A Male Preserve." *Technol. Rev.* **1984,** *87,* 42–43. *See also* McCann, M. "No Parity in Science: Being a Woman Nuclear Physicist." *Sci. People* **1980,** *46,* 9–11;

Chapter 7

1. Teich, M.; Needham, D.M. *A Documentary History of Biochemistry 1770–1940;* Leicester University Press: Leicester, UK, 1992; Ihde, A. J. *The Development of Modern Chemistry;* Harper & Row: New York, 1964.

2. Reil, J. C. "Von der Lebenskraft." *Archiv. Phys.* **1796,** *1,* 8–54.

3. Hein, H. "The Endurance of the Mechanism-Vitalism Controversy." *J. Hist. Biol.* **1972,** *5,* 159–188.

4. Kauffman, G. B.; Choolian, S. H. "Wöhler's Synthesis of Artificial Urea." *J. Chem. Educ.* **1979**, *56*, 197–200.

5. Warren, W. H. "Contemporary Reception of Wöhler's Discovery of the Synthesis of Urea." *J. Chem. Educ.* **1928**, *5*, 1539–1557.

6. Lipman, T. O. "Wöhler's Preparation of Urea and the Fate of Vitalism." *J. Chem. Educ.* **1964**, *41*, 452–458; Brooke, J. H. "Wöhler's Urea, and Its Vital Force?—A Verdict from the Chemists." *Ambix* **1968**, *15*, 84–114.

7. Teich, M. "On the Historical Foundations of Modern Biochemistry." *Clio Med.* **1965**, *1*, 41–57.

8. Kohler, R. E., Jr. "The Enzyme Theory and the Origin of Biochemistry." *Isis* **1973**, *64*, 181–196. © 1973, University of Chicago Press, reprinted by permission.

9. Hofmeister, F. "Die chemische Organisation der Zelle." *Naturwiss. Rundsch.* **1901**, *16*, 581–614.

10. Hopkins, F. G. "The Dynamic Side of Biochemistry" *Rep. Br. Assoc.* **1913**, 652–668.

11. Rossiter, M. W. "'Womens Work' in Science, 1880–1910." *Isis* **1980**, *71*, 381–398; also in Rossiter, M. W. *Women Scientists in America: Struggles and Strategies to 1940;* Johns Hopkins University Press: Baltimore, MD, 1982; pp 53–57.

12. Creese, M. R. S. "British Women of the 19th and Early 20th Centuries Who Contributed to Research in the Chemical Sciences." *Br. J. Hist. Sci.* **1991**, *24*, 275–306.

13. Creese, M. R. S. "British Women of the 19th and Early 20th Centuries," p 296.

14. Needham, J. "Sir F. G. Hopkins' Personal Influence and Characteristics." In *Hopkins and Biochemistry*; Needham, J.; Baldwin, E., Eds.; Cambridge University Press: Cambridge, UK, 1949; p 114.

15. Dixon, M. "Sir F. Gowland Hopkins, O.M., F.R.S." *Nature (London)* **1949**, *160*, 44. Reprinted with permission © 1949 Macmillan Magazines Limited.

16. Rossiter, M. W. "Mendel the Mentor." *J. Chem. Educ.* **1994**, *71*, 215–219.

17. Cited in Rossiter, "Mendel the Mentor." p 218.

18. *See* correspondence files of F. Gowland Hopkins, Hopkins Archives, University of Cambridge.

19. Akeroyd, F. M. "Research Programmes and Empirical Results." *Br. J. Hist. Sci.* **1988**, *39*, 51–58.

20. Reports of these Societies and Clubs were sent to Hopkins, and the items are to be found in Hopkins's archives, Cambridge.

21. Rossiter, "Mendel the Mentor," p 215.

22. Rossiter, "Mendel the Mentor," p 215.

23. Appel, T. A. "Physiology in American Women's Colleges: The Rise and Decline of a Female Subculture." *Isis* **1994**, *85*, 26–56.

24. Kopperl, S. J. "Icie Macy Hoobler: Pioneer Woman Biochemist." *J. Chem. Educ.* **1988**, *65*, 97–98; Williams, H. H. "Icie Gertrude Macy Hoobler (1892–1984). A Biographical Sketch." *J. Nutr.* **1984**, *114*, 1351–1362;

Cavanaugh, M. A. "Icie Gertrude Macy (1892–1984)." In *Women in Chemistry and Physics: A Biobibliographic Sourcebook*; Grinstein, L. S.; Rose, R. K.; Rafailovich, M. H., Eds.; Greenwood Press: Westport, CT, 1993; pp 346–353.

25. Hoobler, I. G. M. with Williams, H. H.; Williams, A. G. *Boundless Horizons, Portrait of a Pioneer Woman Scientist;* Exposition Press: Smithtown, NY, 1982; p 37.

26. Fleck, G. "Mary Lura Sherrill (1888–1968)." In *Women in Chemistry and Physics: A Biobibliographic Sourcebook;* Grinstein, L. S.; Rose, R. K.; Rafailovich, M. H., Eds.; pp 530–537.

27. McCoy, H. N. "Julius Stieglitz, 1867–1937." *J. Am. Chem. Soc.* **1938,** *60,* 3–21 (note: these are special obituary section page numbers).

28. Macy, I. G.; Eckley, J. B. "Sensitiveness of Some Cyanide Reactions." *Proc. Colo. Sci. Soc.* **1918,** *11,* 269.

29. Hoobler, *Boundless Horizons;* p 55.

30. Macy, I. G.; Mendel, L. B. "Comparative Studies on the Physiological Value and Toxicity of Cotton Seed and Some of Its Products." *J. Pharm. Exp. Ther.* **1920,** *16(5),* 345–390.

31. Hoobler, *Boundless Horizons;* p 72.

32. Cavanaugh, M. A. "Agnes Fay Morgan (1884–1968)." In *Women in Chemistry and Physics: A Biobibliographic Sourcebook;* Grinstein, L. S.; Rose, R. K.; Rafailovich, M. H.; pp 434–448.

33. Macy, I. G. "Metabolic and Biochemical changes in Normal Pregnancy." *J. Am. Med. Assoc.* **1958,** *168,* 2265–2271.

34. Macy, I. G.; Mack, H. C. "Implications of Nutrition in the Life Cycle of Woman." *Am. J. Obstet. Gynecol.* **1954,** *68,* 131–150.

35. Macy, I. G.; Kelly, H. G. *Chemical Anthropology: A New Approach to Growth in Children;* University of Chicago Press: Chicago, IL, 1957.

36. Hoobler, *Boundless Horizons;* p 161.

37. Creese M. R. S.; Creese, T. M. "Mary Engle Pennington (1872–1952)." In *Women in Chemistry and Physics: A Biobibliographic Sourcebook;* Grinstein, L. S.; Rose, R. K.; Rafailovich, M. H., Eds.; pp 461–469.

38. Pennington, M. E. "Derivatives of Columbium and Tantalum." *J. Am. Chem. Soc.* **1896,** *18,* 38–67.

39. Pennington, M. E. "Bacterial Growth and Chemical Changes in Milk Kept at Low Temperatures." *J. Biol. Chem.* **1908,** *4,* 353–394.

40. Robinson, L. M. "Regulating What We Eat: Mary Engle Pennington and the Food Research Laboratory." *Agri. Hist.* **1990,** *64,* 143–153. © 1990 by Agricultural History. Reprinted by permission.

41. Pennington, M. E. "Chemical and Bacteriological Study of Fresh Eggs." *J. Biol. Chem.* **1910,** *7,* 109–132.

42. Pennington, M. E.; Clark, E. D. "Refrigeration, Transportation and Conservation of Poultry and Fish Products." *J. Soc. Med.* **1915,** *16,* 272–305.

43. Pennington, M. E. "The Development of a Standard Refrigerator Car." *J. Am. Soc. Refrig. Eng.* **1919–20,** *6,* 1, 10–11.

44. Meites, S. "Willey Glover Denis (1879–1929), Pioneer Woman of Clinical Chemistry." *Clin. Chem.* **1985,** *31,* 774–778.

45. Denis, W. "The Rate of Diffusion of the Inorganic Salts of the Blood into Solutions of Non-Electrolytes and Its Bearing on the Theories of the Immediate Stimulus to the Heart Rhythm." *Am. J. Physiol.* **1906–1907**, *17*, 35–41.

46. Kampmeier, R. H. "Ann Stone Minot (1894–1980): Clinical Chemist and Teacher." *Clin. Chem.* **1986**, *32*, 1602–1609.

47. Denis, W. "The Non-Protein Organic Constituents in the Blood of Marine Fish." *J. Biol. Chem.* **1922**, *54*, 693–700.

48. Folin, O.; Denis, W. "On Phosphotungstic–Phosphomolybdic Compounds as Color Reagents." *J. Biol. Chem.* **1912**, *12*, 239–242.

49. Folin, O.; Denis, W. "New Methods for the Determination of Total Non-Protein Nitrogen, Urea and Ammonia in Blood." *J. Biol. Chem.* **1912**, *11*, 527–536.

50. Most of the information on Brown was obtained from Baldwin, R. S. *The Fungus Fighters: Two Women Scientists and Their Discovery;* Cornell University Press: Ithaca, NY, 1981.

51. Wadsworth, A. B.; Brown, R. "A Specific Antigenic Carbohydrate of Type I Pneumococcus." *J. Immunol.* **1931**, *21*, 243–253.

52. Hazen, E. L.; Reed, F. C. *Laboratory Identification of Pathogenic Fungi Simplified;* Charles C. Thomas: Springfield, IL, 1955.

53. Schatz, A.; Hazen, E. L. "The Distribution of Soil Microorganisms Antagonistic to Fungi Pathogenic to Man." *Mycologia* **1948**, *40*, 461–477.

54. Baldwin, *The Fungus Fighters*, p 59.

55. Brown, R.; Hazen, E. L. "The Activation of Antifungal Extracts of Acinomycetes by Ultrafiltration Through Gradocol Membranes." *Proc. Soc. Exper. Biol. Med.* **1949**, *71*, 454–457.

56. Cited in Baldwin, *The Fungus Fighters*, p 80.

57. Cited in Baldwin, *The Fungus Fighters*, p 191.

58. This section is assembled from Houssay, B.A. "Carl F. and Gerti T. Cori." *Biochem. Biophys. Acta* **1956**, *20*, 11–16.; Miller, J. A. "Gerty Theresa Radnitz Cori (1896–1957)." In *Women in Chemistry and Physics: A Biobibliographic Sourcebook;* Grinstein, L. S.; Rose, R. K.; Rafailovich, M. H., Eds.; pp. 120–127; McGrayne, S. B. *Nobel Prize Women in Science: Their Lives, Struggles and Momentous Discoveries;* Birch Lane Press: New York, 1992, pp 93–116. See also Cohn, M. "Carl and Gerti Cori: A Personal Recollection." In *Creative Couples in the Sciences;* Pycior, H. M.; Slack, N. G.; Abir-Am, P. G., Eds.; Rutgers University Press: New Brunswick, NJ, 1996; pp 72–84.

59. Radnitz, G; Cori, K. "Über den Gehalt des Menschlichen Blutserums an Komplement und Normalambozeptor für Hammelblutkörperchen." *Z. Immun.* **1920**, *29*, 445–462.

60. Cori, G. "Comparison of the Sensitiveness of Different Organs of the Mouse Towards X-ray." *J. Cancer Res.* **1924**, *8*, 522–524.

61. Cori, G. "The Insulin Content of Cancer Tissue." *J. Cancer Res.* **1925**, *9*, 408–410.

62. McGrayne, *Nobel Prize Women in Science*, p 102.

63. *See,* for example, Cori, G. "Carbohydrate Changes during Anaerobiosis of

Mammalian Tissue." *J. Biol. Chem.* **1932**, *96*, 259–269; Cori, C. F.; Cori G. "A Comparison of Total Carbohydrate and Glycogen Content of Mammalian Muscle." *J. Biol. Chem.* **1933**, *100*, 323–332.

64. McGrayne, *Nobel Prize Women in Science;* p 103.

65. Cori, G.; Illingworth, B.; Keller, P. J. "Muscle Phosphorylase." *Meth. Enzymol.* **1955**, *1*, 200–205.

66. This section is assembled from Goodman, M. "Gertrude Belle Elion (1918–)." In *Women in Chemistry and Physics: A Biobibliographic Sourcebook;* Grinstein, L. S.; Rose, R. K.; Rafailovich, M. H., Eds.; pp 169–179; McGrayne, S. B. *Nobel Prize Women in Science: Their Lives, Struggles and Momentous Discoveries;* pp 280–303.

67. McGrayne, *Nobel Prize Women in Science,* p 286.

68. Cited in McGrayne, *Nobel Prize Women in Science,* p 286.

69. Elion, G. B. "Historical Background of 6-Mercaptopurine." *Toxic. Ind. Health* **1986**, *2*, 1–9.

70. Elion, G. B.; Hitchins, G. H. "Thiopurines as Inhibitors of the Immune Response." *Biochem. Aspects Antimetab. Drug Hydroxylation* **1969**, *16*, 1–10.

71. Elion, G. B.; Rundles, R. W.; Hitchins, G. H. "Allopurinol in the Treatment of Gout and Secondary Hyperuricemia." *Bull. Rheum. Dis.* **1966**, *16*, 400–403.

72. Cited in McGrayne, *Nobel Prize Women in Science;* p 299.

73. Elion, G. B. "The Purine Path to Chemotherapy." *Science (Washington, D.C.)* **1989**, *244*, 41–47.

74. Selection 120 in Teich, M.; Needham, D. M. *A Documentary History of Biochemistry 1770–1940*, pp 314–316.

75. We thank Ann Phillips, Archivist, Newnham College, Cambridge for biographical information.

76. Vardy, W. I. *King Edward VI High School for Girls Birmingham 1883–1925;* Ernest Benn: London, 1928. Among the graduates were the prominent chemist, Ida Smedley MacLean, mentioned in Chapter 4 of this book, and one of Dorothy Hodgkin's students, the crystallographer Jenny Pickworth Glusker. For additional information on MacLean, *see* Whiteley, M. A. "Ida Smedley MacLean. 1877–1944." *J. Chem. Soc.* **1946**, 65–67; and for information on Glusker, *see* Rose, R. K.; Glusker, D. L.; "Jenny Pickworth Glusker (1931–)." In *Women in Chemistry and Physics: A Biobibliographic Sourcebook;* Grinstein, L. S.; Rose, R. K.; Rafailovich, M. H., Eds.; pp 207–217.

77. Willcock, E. G. "Radium and Animals." *Nature (London)* **1903**, *69*, 55.

78. Cited in Stephenson, M. "Sir F. G. Hopkins' Teaching and Scientific Influence." In *Hopkins & Biochemistry 1861–1947;* Needham, J.; Baldwin, E., Eds.; W. Heffer: Cambridge, UK, 1949; p 34.

79. Willcock, E. G.; Hopkins, F. G. "The Importance of Individual Amino Acids in Metabolism." *J. Physiol.* **1907**, *35*, 88–102.

80. Information on Menten was found in Ainley, M. G.; Crossfield, T. "Canadian Women's Contributions to Chemistry, 1900–1970." *Can. Chem. News*

1994, *46(4),* 16–18; Fetterman, G. H. "Maud L. Menten (1879–1960)." *Perspect. Pediatr. Pathol.* **1984,** *1(1),* 5–7.

81. Menten, M. L. "Experiments on the Influence of Radium Bromide on a Carcinomatous Tumor of the Rat." *Monogr. Rockefeller Inst. Med. Res.* **1910,** *1,* 73–80.

82. Michaelis, L.; Menten, M. L. "Die Kinetik der Invertinwirkung." *Biochem. Z.* **1913,** *49,* 333–369.

83. Fetterman, "Maud L. Menten (1879–1960)," p 6.

84. Andersch, M. A.; Wilson, D. A.; Menten, M. L. "Sedimentation Constants and Electrophoretic Mobilities of Adult and Fetal Carbonylhemoglobin." *J. Biol. Chem.* **1944,** *153,* 301–305.

85. Menten, M. L.; Junge, J.; Green, M. H. "Distribution of Alkaline Phosphatase in Kidney Following the Use of Histochemical Azo Dye Test." *Proc. Soc. Exp. Biol. Med.* **1944,** *57,* 82–86.

86. Fetterman, "Maud L. Menten (1879–1960)," p 6.

87. Stock, A. H.; Carpenter, A. M. "Prof. Maud Menten, Obituary." *Nature (London)* **1961,** *189,* 965. Reprinted with permission © 1961 Macmillan Magazines Limited.

88. This section is assembled from: Abir-Am, P. G. "Dorothy Maud Wrinch (1894–1976)." In *Women in Chemistry and Physics: A Biobibliographic Sourcebook;* Grinstein, L. S.; Rose, R. K.; Rafailovich, M. H., Eds.; pp 605–612; Julian, M. M. "Dorothy Wrinch and the Search for the Structure of Proteins." *J. Chem. Educ.* **1984,** *61,* 890–892; Abir-Am, P. G. "Synergy or Clash: Disciplinary and Marital Strategies in the Career of Mathematical Biologist Dorothy Wrinch." In *Uneasy Careers and Intimate Lives: Women in Science, 1789–1979;* Abir-Am, P. G.; Outram, D.; Eds.; Rutgers University Press: New Brunswick, NJ, 1987; pp 239–280.

89. Abir-Am, P. G. "The Biotheoretical Gathering, Trans-Disciplinary Authority and the Incipient Legitimation of Molecular Biology in the 1930s: New Perspective on the Historical Sociology of Science." *Hist. Sci.* **1987,** *25,* 64.

90. Ayling, J. *The Retreat from Parenthood;* Kegan Paul, Trench, Trubner & Co.: London, 1930.

91. Cited in Julian, "Dorothy Wrinch and the Search for the Structure of Proteins," p 890.

92. Wrinch, D. M.; Langmuir, I. "Nature of the Cyclol Bond." *Nature (London)* **1939,** *143,* 49.

93. Hodgkin, D. C. "Obituary." *Nature (London)* **1976,** *260,* 564. Reprinted with permission © 1976 Macmillan Magazines Limited.

94. Cited in Selection 134 in Teich, M.; Needham, D.M. *A Documentary History of Biochemistry 1770–1940;* p 349.

95. Martin, R. B. "Dorothy Wrinch and the Structure of Proteins." *J. Chem. Educ.* **1987,** *64,* 1069.

96. Hodgkin, "Obituary," 564.

97. Wrinch, D. *Fourier Transforms and Structure Factors;* American Society for X-ray and Electron Diffraction: Cambridge, MA, 1946.

98. Roscher, N. M.; Nguyen, C. K. "Helen M. Dyer, A Pioneer in Cancer Research." *J. Chem. Educ.* **1986,** *63,* 253–255; Roscher, N. M.; Nguyen,

C. K. "Helen Marie Dyer (1895–)." In *Women in Chemistry and Physics: A Biobibliographic Sourcebook;* Grinstein, L. S.; Rose, R. K.; Rafailovich, M. H., Eds.; pp 162–168.

99. Roscher, N. M.; Nguyen, C. K. "Pauline Gracia Beery Mack (1891–1974)." In *Women in Chemistry and Physics: A Biobibliographic Sourcebook;* Grinstein, L. S.; Rose, R. K.; Rafailovich, M. H., Eds.; pp 337–345.

100. Kohler, R. E. "Innovation in Normal Science: Bacterial Physiology." *Isis* **1985,** *76,* 162–181; Robertson, M. "Marjory Stephenson 1885–1948." *R. Soc. (Great Br.), Obituary Notices Fellows,* no. 6, 563–577; Mason, J. "Marjory Stephenson 1885–1948." In *Cambridge Women: Twelve Portraits;* Shils, E.; Blacker, C., Eds.; Cambridge University Press: Cambridge, UK, 1996; pp 113–135.

101. Kathleen Lonsdale was the first woman elected to the Royal Society. See Mason, J. "The Women Fellows' Jubilee." *Notes Records R. Soc. London* **1995,** *49(1),* 125–140.

102. Needham, D. "Women in Cambridge Biochemistry." In *Women Scientists: The Road to Liberation;* Richter, D., Ed.; Macmillan: London; p 161.

Chapter 8

1. Rhees, D. J. "The Chemists' War: The Impact of World War I on the American Chemical Profession." *Bull. Hist. Chem.* **1992/1993,** *13/14,* 40–47.

2. Heilbron, J. L. *H.G.J. Moseley; The Life and Letters of an English Physicist, 1887–1915;* University of California Press: Berkeley, CA, 1974.

3. Woollacott, A. *On Her Their Lives Depend: Munition Workers in the Great War;* University of California Press: Berkeley, CA, 1994.

4. Rayner-Canham, M. F.; Rayner-Canham, G. W. "The Gretna Garrison." *Chem. Br.* **1996,** *32(3),* 37–41.

5. *Home Office Report: Substitution of Women in Non-munitions Factories During the War,* His Majesty's Stationery Office, London, 1919.

6. Letter from Fred K. Knowles, Faculty of Metallurgy, University of Sheffield, in the Women's War Work File, Imperial War Museum, London.

7. Rayner-Canham, M. F.; Rayner-Canham, G. W. "Ada Hitchins: Research Assistant to Frederick Soddy." In *A Devotion to Their Science: Pioneer Women of Radioactivity;* Rayner-Canham, M. F.; Rayner-Canham, G. W., Eds.; Chemical Heritage Foundation: Philadelphia, PA, and McGill-Queen's University Press: Montreal, PQ, 1997; pp 152–155.

8. File 201, Papers of Professor Frederick Soddy, Bodleian Library, Oxford.

9. Haynes, W. *American Chemical Industry, The World War I Period: 1912–1922;* van Nostrand: New York, 1945.

10. Shmurak, C. B.; Handler, B. "'Castle of Science': Mount Holyoke College and the Preparation of Women in Chemistry, 1837–1941." *Hist. Educ. Quart.* **1992** (Fall), *32,* 315–342.

11. Rayner-Canham, M. F.; Rayner-Canham, G. W. "A Chemist of Some Repute." *Chem. Br.* **1993,** *29,* 206–208.

12. John Sichel, personal communication, 31 August 1993; Dainton, F. S. "Reputable memories." *Chem. Br.* **1993**, *29*, 573.

13. Smithalls Collection, University of Leeds.

14. Hall, J. A.; Jaques, A.; Leslie, M. S. "Nitric Acid Absorption Towers" *J. Soc. Chem. Ind. Rev.* **1922**, *41*, 285–293.

15. Edward S.F. Rogans, personal communication, 31 August 1993.

16. Rogans, E. S. F. "Reputable Memories." *Chem. Br.* **1993**, *29*, 573.

17. These unpublished reports were entitled "Experiments in the design of absorption towers for nitrous fumes" (October 1916) and "Experiments in the design of absorption towers for nitrous fumes: Supplementary Report" (8 November 1916), Archives Section, Edward Boyle Library, University of Leeds. We thank Peter Towse, University of Leeds, for this information.

18. We thank J. Mary D. Forster, University Archive, University of Leeds for a copy of this report.

19. "Obituary: Arthur Hamilton Burr." *Inst. Chem. Great Br. Ireland, J. Proc.* **1934**, 68.

20. Burr, M. S. (née Leslie) *The Alkaline Earth Metals; A Textbook of Inorganic Chemistry;* part 1; Friend, J. N., Ed.; Charles Griffin: London, 1925; Vol. 3; Gregory, J. C.; Burr, M. S. (née Leslie), *Beryllium and Its Congeners; A Textbook of Inorganic Chemistry;* part 2; Friend, J. N., Ed.; Charles Griffin: London, 1926; Vol. 3.

21. Rossiter, M. W. *Women Scientists in America: Struggles and Strategies to 1940;* Johns Hopkins University Press: Baltimore, MD, 1982; p 71.

22. We thank Frances Boyle, Royal Society of Chemistry Information Services, for this information.

23. *Yorkshire Post,* 6 July 1937, 5.

24. Dawson, H. M. "May Sybil Burr. 1887–1937." *J. Chem. Soc.* **1938**, 151–152.

25. Wall, F. E. "Reflections on 70 Years of Chemistry." *The Chemist* **1986**, *63*, 16–17.

26. Alvarez, R. "Radiation Workers: The Dark Side of Romancing the Atom." *Sci. People* **1986** (Mar/Apr), *18*, 6–11. The tragic story of the denial of radium-related disease among the young women who applied radium-based paint to watch dials is told in Clark, C. "Physicians, Reformers and Occupational Disease: The Discovery of Radium Poisoning." *Women Health* **1987**, *12*, 147–167.

27. Wall, F. E. "Training in Chemistry for the Cosmetics Industry." *J. Chem. Educ.* **1936**, *13*, 432–436; "Cosmetics—A Fertile Field for Chemical Research." *J. Chem. Educ.* **1940**, *17*, 77–81; "Cosmetics." *J. Chem. Educ.* **1942**, *19*, 435–440.

28. Wall, F. E. *The Science of Beautistry: Official Textbook;* National School of Cosmeticians: New York, 1932; Wall, F. E. *The Principles and Practice of Beauty Culture,* 1st. ed.; Keystone: New York, 1941. (4th ed., 1961).

29. "Florence E. Wall, Authority on Cosmetics, Dies at Age 95," *New York Times,* October 12, section 4, p 27.

30. Wall, F. E. "Bleaches, Hair Colorings, and Dye Removers." In *Cosmetic Science and Technology,* 2nd. ed.; Balsam, M. S., Ed.; Wiley: New York, 1972; pp 279–343.

31. Rossiter, M. W. *Women Scientists in America: Struggles and Strategies to 1940;* Johns Hopkins University Press: Baltimore, MD, 1982; p 118.

32. Anon., "Prospects of Employment for Women Science Graduates III: Industrial Research Laboratories." *J. Careers* **1938**, *17*, 289–296. The U.S. Federal Government hired women chemists, but almost exclusively for the Bureau of Home Economics. *See* Worner, R. K. "Opportunities for Women Chemists in Washington." *J. Chem. Educ.* **1939**, *16*, 583–585.

33. Rossiter, M. W. *Women Scientists in America: Struggles and Strategies to 1940,* p 289.

34. Anon., "Prospects of Employment for Women Science Graduates Part I of a Survey of Opportunities in Government, Industrial and Other Research Laboratories." *J. Careers* **1938**, *17*, 88–93.

35. This section is largely complied from Davis, K. A. "Katharine Blodgett and Thin Films." *J. Chem. Educ.* **1984**, *61*, 437–439.

36. Langmuir, I. "The Constitution and Fundamental Properties of Solids and Liquids: II Liquids." *J. Am. Chem. Soc.* **1917**, *39*, 1848–1906. In this paper, Langmuir acknowledges the earlier contributions of Agnes Pockels (*see* Chapter 3 of this book).

37. Langmuir, I. "The Mechanism of the Surface Phenomena of Flotation." *Trans. Faraday Soc.* **1920**, *15*, 62–74.

38. Blodgett, K. "Films Built by Depositing Successive Monomolecular Layers on a Solid Surface." *J. Am. Chem. Soc.* **1935**, *57*, 1007–1022.

39. Quoted in Yost, E. *American Women of Science;* Frederick A. Stokes: Philadelphia, PA, 1943, p 207.

40. Blodgett, K. "Use of Interference To Extinguish Reflection of Light from Glass." *Phys. Rev.* **1939**, *55*, 391–404.

41. Gleiser, M. "The Garvan Women." *J. Chem. Educ.* **1985**, *62*, 1056–1068.

42. Gaines, G. L. "In Memoriam: Katharine Burr Blodgett 1898–1979." *Thin Solid Films* **1980**, *68*, vii–viii; Schaefer, V. J.; Gaines, G. L. "Obituary: Katharine Burr Blodgett 1898–1979." *J. Colloid Interface Sci.* **1980**, *76*, 269–270.

43. Suits, C. G. "Seventy-five Years of Research in General Electric." *Science (Washington, D.C.)* **1953**, *118*, 451–456; Wise, G. "Ionists in Industry: Physical Chemistry at General Electric, 1900–1916." *Isis* **1983**, *74*, 7–21.

44. Finley, K. T.; Siegel, P. J. "Katharine Burr Blodgett (1898–1979)." In *Women in Chemistry and Physics: A Biobibliographic Sourcebook;* Grinstein, L. S.; Rose, R. K.; Rafailovich, M. H., Eds.; Greenwood Press: Westport, CT, 1993; pp 65–71.

45. Petty, M. *Langmuir–Blodgett Films: An Introduction;* Cambridge University Press: Cambridge, UK, 1996.

46. This section is compiled from Bramley, R. "Kathleen Culhane Lathbury." *Chem. Br.* **1991**, *27*, 428–431; from Bramley, R. "Mrs. Kathleen Lathbury B.Sc., C.Chem., F.R.S.C.." unpublished.

47. Bingham, C. *The History of Royal Holloway College 1886–1986;* Constable: London, 1987.

48. Culhane, K. "Variations in the Serum-Calcium of Rabbits." *Biochem. J.* **1927**, *21*, 1015–1023.

49. Cited in Bramley, R. "Kathleen Culhane Lathbury," *Chem. Br.* 428.
50. Culhane, K. "The Use of Rabbits in Insulin Assay." *J. Pharmacol.* **1928**, *1*, 517–533.
51. Cited in Bramley, R. "Kathleen Culhane Lathbury," p 430.
52. Beddoe, D. *Women Between the Wars, 1918–1939: Back to Home and Duty;* Pandora: London, 1989; p 82.
53. Cited in Bramley, R. "Kathleen Culhane Lathbury," p 430.
54. Landis, W. S. "Women Chemists in Industry." *J. Chem. Educ.* **1939**, *16*, 577–579.
55. Rossiter, M. W. "Chemical Librarianship: A Kind of Women's Work' in America." *Ambix* **1996**, *43*, 46–58.
56. Cortelyou, E. "Counseling the Woman Chemistry Major." *J. Chem. Educ.* **1955**, *32*, 196–197.
57. "Stephanie Kwolek." Unpublished biography, DuPont Advanced Fibers Systems, Wilmington, DE, 1997.
58. Cavanaugh, M. A. "Mary Lowe Good (1931–)." In *Women in Chemistry and Physics: A Biobibliographic Sourcebook;* Grinstein, L. S.; Rose, R. K.; Rafailovich, M. H., Eds.; Greenwood Press: Westport, CT, 1993; pp 218–229.

Chapter 9

1. Creese, M. R. S.; Creese, T. M. "Charlotte Roberts and Her Textbook on Stereochemistry." *Bull. Hist. Chem.* **1994**, *15/16*, 31–36. The earliest prominant woman chemist educator in the United States was Almira Hart Lincoln Phelps (1793–1884). *See* Weeks, M. A.; Dains, F. B. "Mrs. A. H. Lincoln Phelps and Her Services to Chemical Education." *J. Chem. Educ.* **1937**, *14*, 53–57.
2. Roberts, C. F. *The Development and Present Aspects of Stereochemistry;* D. C. Heath: Boston, MA, 1896.
3. The section on Cremer is summarized from: Miller, J. A. "Erika Cremer (1900–)." In *Women in Chemistry and Physics: A Biobibliographic Sourcebook;* Grinstein, L. S.; Rose, R. K.; Rafailovich, M. H., Eds.; Greenwood Press: Westport, CT, 1993; pp 128–135.
4. Cremer, E. "Über die reaktion zwischen Chlor, Wasserstoff und Sauerstoff im Licht." *Z. Phys. Chem.* **1927**, *128*, 285–317.
5 Frevert, U. *Women in German History;* Berg: Oxford, England, 1989, p. 197; Koonz, C. "Nazi Women Before 1933: Rebels Against Emancipation." *Soc. Sci. Quart.* **1976**, *56*, 553–563.
6. *A History of Analytical Chemistry;* Laitenen, H. A.; Ewing, G. W., Eds.; Division of Analytical Chemistry of the American Chemical Society: Washington, DC, 1977; p 297.
7. Cremer, E. "How We Started To Work in Gas Adsorption Chromatography." *Chromatographia* **1976**, *9*, 364–366.
8. This paper was finally published in 1976. *See* Ettre, L. S. "Dr. Cremer's First, Unpublished Paper on Gas Chromatography." *Chromatographia* **1976**, *9*, 363.
9. These papers included Cremer, E.; Müller, R. "Separation and Determination of Small Quantities of Gases by Chromatography." *Mikrochem.Mikrochem. Acta* **1951**, *36/37*, 553–560.

10. "M. S. Tswett Chromatography Medal Awarded to Erica Cremer." *Chem. Eng. News* **1974**, *52(45)*, 26.

11. Bonn, G. "Foreword." *Chromatographia* **1990**, *30*, 469; Bobleter, O. "Professor Erika Cremer—A Pioneer in Gas Chromatography." *Chromatographia* **1990**, *30*, 471–476.

12. Most of the biographical material is obtained from Chalmers, R. A. "A Mastery of Microanalysis." *Chem. Br.* **1993**, *29*, 492–494.

13. Miller, C. C. "The Stokes–Einstein Law for Diffusion in Solution." *Proc. R. Soc.* **1924**, *106A*, 724–749.

14. Miller, C. C. "8-Hydroxyquinoline as a Reagent for the Determination of Magnesium, Especially in Carbonate and Silicate Rocks." *J. Chem. Soc.* **1940**, 656–659.

15. Chalmers, "A Mastery of Microanalysis," p 494.

16. Carr, E. P. "Chemical Education in American Institutions: Mount Holyoke College." *J. Chem. Educ.* **1948**, *25*, 11–15.

17. Handler B. S.; Shmurak, C. B. "Rigor, Resolve, Religion: Mary Lyon and Science Education." *Teach. Educ.* **1991**, *3*, 137–142.

18. Shmurak, C. B.; Handler, B. S. "Lydia Shattuck: A Streak of the Modern." *Teach. Educ.* **1991**, *3*, 127–129.

19. Rossiter, M. W. *Women Scientists in America: Struggles and Strategies to 1940;* Johns Hopkins University Press: Baltimore, MD, 1982; p 78.

20. Shmurak, C. B.; Handler, B. S. "Castle of Science: Mount Holyoke College and the Preparation of Women in Chemistry, 1837–1941." *Hist. Educ. Quart.* **1992**, *32*, 315–342.

21. Cited in: Shmurak, C. B.; Handler, B. "'Castle of Science': Mount Holyoke College and the Preparation of Women in Chemistry, 1837–1941," p 315.

22. The biographical details on Carr have been assembled from Banville, D. L. "Emma Perry Carr (1880–1972)." In *Women in Chemistry and Physics: A Biobibliographic Sourcebook;* Grinstein, L. S.; Rose, R. K.; Rafailovich, M. H., Eds.; pp 77–84; Shmurak, C. B. "Emma Perry Carr: The Spectrum of Life." *Ambix* **1994** (July), *41*, 75–86; Jennings, B. H. "The Professional Life of Emma Perry Carr." *J. Chem. Educ.* **1986**, *63*, 923–927.

23. Carr, E. P. "Research in a Liberal Arts College." *J. Chem. Educ.* **1957**, *34*, 467–470.

24. Shmurak, C. B. "Emma Perry Carr: The Spectrum of Life," p 78.

25. Carr, E. P.; Burt, C. P. "The Absorption Spectra of Some Derivatives of Cyclopropane." *J. Am. Chem. Soc.* **1918**, *40*, 1590–1600.

26. Cited in Shmurak, C. B.; Handler, B. "'Castle of Science': Mount Holyoke College and the Preparation of Women in Chemistry, 1837–1941," p 337.

27. Cited in Jennings, B. H. "The Professional Life of Emma Perry Carr," p 923.

28. Fleck, G. "Mary Lura Sherrill (1888–1968)." In *Women in Chemistry and Physics: A Biobibliographic Sourcebook;* Grinstein, L. S.; Rose, R. K.; Rafailovich, M. H., Eds.; pp 530–537.

29. This work is summarized in Carr, E. P. "Electronic Transitions in the Simple Unsaturated Hydrocarbons." *Chem. Rev.* **1947,** *41,* 293–299.

30. For details on the Garvan award and the recipients, *see* Gleiser, M. "The Garvan Women." *J. Chem. Educ.,* **1985,** *62,* 1065–1068; Roscher, N. M. "Women Chemists." *CHEMTECH* **1976,** 738–743.

31. "In Memoriam: Emma Perry Carr." *Mount Holyoke Alumni Quart.* **1972** (Spring), 23–25.

32. Another author of organic texts is Marjorie Caserio. *See* Goldwhite, H. "Marjorie Constance Beckett Caserio (1929–)." In *Women in Chemistry and Physics: A Biobibliographic Sourcebook;* Grinstein, L. S.; Rose, R. K.; Rafailovich, M. H., Eds.; pp 85–93.

33. Fieser's life and work is summarized from Pramer, S. "Mary Fieser: A Transitional Figure in the History of Women." *J. Chem. Educ.* **1985,** *62,* 186–191. *See also* Schulz, W. "Harvard's Fieser Laboratory Honors Legacy of Husband-and-Wife Team." *Chem. Eng. News* **1996,** April 29, 75–76.

34. Pramer, "Mary Fieser," p 186.

35. Pramer, "Mary Fieser," p 187.

36. Her first publication was Fieser, L. F.; Peters, M. A. "The Potentials and the Decomposition Reactions of Ortho Quinones in Acid Solution." *J. Am. Chem. Soc.* **1931,** *53,* 793–805.

37. Pramer, "Mary Fieser," p 188.

38. Fieser, L. F. "Lapinone, a New Antimalarial." *Ciencia* **1951,** *11,* 65–74.

39. Fieser, L. F.; Fieser, M. *Organic Chemistry;* Reinhold: New York, 1946.

40. Fieser, L. F.; Fieser, M. *Introduction to Organic Chemistry;* Reinhold: New York, 1946.

41. Fieser, L. F.; Fieser, M. *Basic Organic Chemistry;* D. C. Heath: Boston, MA, 1958.

42. Fieser, L. F.; Fieser, M. *Style Guide for Chemists;* Reinhold: New York, 1960.

43. Fieser, L. F.; Fieser, M. *Advanced Organic Chemistry;* Reinhold, New York, 1961.

44. Fieser, L. F.; Fieser, M. *Topics in Organic Chemistry;* Reinhold: New York, 1963.

45. Fieser and Fieser's Staff, *Fieser and Fieser's Reagents for Organic Synthesis;* 17 vol.; Wiley: New York, 1994.

46. Freudenthal, G. "Hélène Metzger: Eléments de biographie." In *Études sur Hélène Metzger;* Freudenthal, G., Ed.; E. J. Brill: Leiden, Germany, 1990; pp 197–208; Delorme, S. "Metzger, Hélène." In *Dictionary of Scientific Biography, vol. IX;* Gillispie, C. C., Ed.; Charles Scribner: New York, 1974; pp 340–342.

47. H. Metzger to G. Sarton, 22 April 1926, cited in Freudenthal, "Hélène Metzger," p 198.

48. Golinski, J. "Hélène Metzger and the Interpretation of Seventeenth Century Chemistry," *Hist. Sci.* **1987,** *25,* 85–97.

49. The latest edition is Metzger, H. *Chemistry,* translated and annotated by Michael, C. V.; Locust Hill Press: West Cornwall, CT, 1991.

50. Bensaude-Vincent, B. "Hélène Metzger's *la chimie*. A Popular Treatise," *Hist. Sci.* **1987**, *25*, 71–84.

51. *See,* for example, the following commentaries: Freudenthal, G., Ed.; *Études sur Hélène Metzger;* E. J. Brill: Leiden, Germany, 1990; Boas, M. "Notice Nécrologique: Hélène Metzger (1889–1944)." *Arch. Int. Hist.Sci.* **1955**, *8*, 433–434; Christie, J. R. R. "Narrative and Rhetoric in Hélène Metzger's Historiography of 18th-Century Chemistry." *Hist. Sci.* **1987**, *25*, 99–109; Freudenthal, G. "The Hermeneutical Status of the History of Science: The Views of Hélène Metzger." *Organon: Int. Rev. (Polish)* **1986–87**, 22–23 and 33–39.

52. Delorme, S. "Metzger, Hélène," p 342.

53. Sarton, G. "Notes and Correspondence: Hélène Metzger (1889–1944)." *Isis* **1946**, *36*, 133; Freudenthal, "Hélène Metzger," p 198

Chapter 10

1. Graham, P. A. "Expansion and Exclusion: A History of Women in American Higher Education." *Signs* **1978**, *3*, 759–773. © 1978, University of Chicago Press, reprinted by permission.

2. By contrast, Fox Keller refers to the mid-twentieth century as "the nadir of the history of women in science" adding that by the 1950s "women scientists had effectively disappeared from American science." *See* Fox Keller, E. "The Gender/Science System: or, Is Sex to Gender as Nature Is To Science?" *Hypatia* **1987**, *2*, 40.

3. Nicolson, M. "The Rights and Privileges Pertaining Thereto." *J. Am. Assoc. Univ. Women* **1938**, *31(3)*, 136. *See also* Dyhouse, C. "The British Federation of University Women and the Status of Women in Universities, 1907–1939." *Women's Hist. Rev.* **1995**, *4*, 465–485.

4. Bolton, H. I. "Women in Science." *Pop. Sci. Month.* **1898**, *53*, 506–511.

5. Vetter, B. "Changing Patterns of Recruitment and Employment." In *Women in Scientific and Engineering Professions;* Hass, V. B.; Perucci, C. C., Eds.; University of Michigan Press: Ann Arbor, MI, 1984; p 59.

6. Rayner-Canham, M. F.; Rayner-Canham, G. W. "Women in Chemistry: Participation During the Early Twentieth Century." *J. Chem. Educ.* **1996**, *73*, 203–205.

7. Everett, K. G.; DeLoach, W. S. "Chemistry Doctorates Awarded to Women in the United States: A Historical Perspective." *J. Chem. Educ.* **1991**, *68*, 545–547.

8. Roberts, E. *Women's Work 1840–1940;* Macmillan Education: London, 1988; p 67.

9. Strong-Boag, V. *The New Day Recalled: Lives of Girls and Women in English Canada, 1919–1939;* Copp Clark Pittman: Toronto, ON, 1988; p 25.

10. Cited in de Grazia, V. *How Fascism Ruled Women: Italy 1922–1945;* University of California Press: Berkeley, CA, 1992; p 161.

11. Faludi, S. *Backlash;* Doubleday: New York, 1991; p 50. Reprinted by permission of Crown Publishers, Inc., © 1991. Kent has discussed the parallel decline of feminism in Britain. *See* Kent, S. K. "The Politics of Sexual

Difference: World War I and the Demise of British Feminism." *J. Br. Stud.* **1988**, *27*, 232–253.

12. Freedman, E. "The New Woman: Changing Attitudes of Women in the 1920's." *J. Am. Hist.* **1974**, *61*, 372–393; Hummer, P. *The Decade of Elusive Promise: Professional Women in the United States, 1920–1930;* UMI Research Press: Ann Arbor, MI, 1979.

13. Antler, J. "'After College, What?': New Graduates and the Family Claim." *Am. Quart.* **1980** (Fall), 409–434; Kirejczyk, M. "Vrouwen kozen exact; studie en beroepsuitoefening rond de eeuwwisseling." *Gewina* **1993**, *16*, 234–247.

14. *See*, for example: Ainley, M. G. "'Women's Work' in Canadian Chemistry." *Can. Woman Stud.* **1993**, *13(2)*, 43–46; Carter, S. B. "Academic Women Revisited: An Empirical Study of Changing Patterns in Women's Employment as College and University Faculty, 1890–1960." *J. Soc. Hist.* **1981**, *14*, 675–699.

15. Pauwels, J. R. *Women, Nazis, and Universities: Female University Students in the Third Reich, 1933–1945;* Greenwood Press: Westport, CT, 1984; p 33.

16. Reif, F. "The Competitive World of the Pure Scientist." *Science (Washington, D.C.)* **1961**, *134*, 1957–1962. *See also* Capshew, J. H.; Rader, K. A. "Big Science: Price to the Present." *Osiris* **1992**, *7*, 3–25.

17. Sassen has discussed the data that show a climate of competition arouses anxiety among women far more than men. She argues that rather than training women for competion, institutions should adapt to more collaborative styles of operation. *See* Sassen, G. "Success Anxiety in Women: A Constructivist Interpretation of its Source and Significance." *Harvard Educ. Rev.* **1980**, *50*, 13–24.

18. Freeman, J. *A Passion for Physics: The Story of a Woman Physicist;* Adam Hilger: Bristol, UK, 1991; p 115.

19. Rossiter, M. *Women Scientists in America: Before Affirmative Action 1940–1972;* Johns Hopkins University Press: Baltimore, MD, 1995; p 11.

20. Cited in Rossiter, M. *Women Scientists in America: Before Affirmative Action 1940–1972,* p 18.

21. Keubel, A. "Women in Chemistry." *J. Chem. Educ.* **1943**, *20*, 248–249.

22. Woodford, L. "Trends in the Industrial Employment of Women Chemists." *J. Chem. Educ.* **1945**, *22*, 236–238.

23. Woodford, L. "Opportunities for Women in Chemistry." *J. Chem. Educ.* **1942**, *19*, 536–538.

24. "Women Chemists in New York." *Chem. Eng. News* **1944**, *22*, 1673.

25. Horsey, E. F.; Price, D. "Science out of Petticoats." *J. Am. Assoc. Univ. Women* **1946**, *40*, 13–16.

26. Carr, E. P. "Letter to Headquarters: Those Petticoats in Science." *J. Am. Assoc. Univ. Women* **1947**, *40*, 107.

27. Snell, C. "Women as Professional Chemists." *J. Chem. Educ.* **1948**, *25*, 450–453.

28. Rossiter, M. *Women Scientists in America: Before Affirmative Action 1940–1972,* p 49.

29. Manning, K. R. "Roger Arliner Young, Scientist." *Sage* **1989**, *6*, 3–7.

30. Howes, R. H. "Chien-Shiung Wu (1912–)." In *Women in Chemistry and Physics: A Biobibliographic Sourcebook;* Grinstein, L. S.; Rose, R. K.; Rafailovich, M. H., Eds.; Greenwood Press: Westport, CT, 1993; pp 613–625.

31. Sammons, V. O. *Blacks in Science and Medicine;* Hemisphere: New York, 1990; p 120.

32. Sammons, *Blacks in Science and Medicine,* p 67.

33. Mayes, V. "Lee Lorch at Fisk: A Tribute." *Am. Math. Month.* **1981,** *88,* 708–711.

34. Hubbard, R. "Facts and Feminism: Thoughts on the Masculinity of Natural Science." *Sci. People* **1986** (Mar/Apr), *18,* 16–20, 26. The discovery of pulsar stars provides the classic example of the claim of graduate students for recognition of their contribution. *See* Wade, N. "Discovery of Pulsars: A Graduate Student's Story." *Science (Washington, D.C.)* **1975,** *189,* 358–364.

35. Hubbard, R. "Facts and Feminism," p 20.

36. Brennan, M. "1962 Garvan Medalist Celebrates 100th Birthday." *Chem. Eng. News* **1995,** *73* (May 29), 53.

37. Rayner-Canham, M. F.; Rayner-Canham, G. W. "Elizabeth Róna: The Polonium Woman." In *A Devotion to Their Science: Pioneer Women of Radioactivity;* McGill–Queen's University Press: Montreal, Canada, 1997; pp 209–216.

38. Eva Ramstedt to Marie Curie, 11 August, 1925, Bibliothèque Nationale. For a biographical account of Ramstedt *see* Rayner-Canham, M. F.; Rayner-Canham, G. W. ". . .And Some Other Women of the French Group." In *A Devotion to Their Science: Pioneer Women of Radioactivity;* Rayner-Canham, M. F.; Rayner-Canham, G. W., Eds.; Chemical Heritage Foundation: Philadelphia, PA, and McGill–Queen's University Press: Montreal, PQ, 1997; pp 125–126.

39. Cited in Barnard, J. *Academic Women;* Pennsylvania State University Press: University Park, PA, 1964; p 207.

40. Cited in Apter, T. *Professional Progress: Why Women Still Don't Have Wives;* Macmillan: Basingstoke, UK, 1993; p 1.

41. Etzkowitz, H; Kemelgor, C.; Neuschatz, M.; Uzzi, B.: Alonzo, J. "The Paradox of Critical Mass for Women in Science." *Science (Washington, D.C.)* **1994,** *266,* 51–54.

Appendix: The Garvan Medalists

Francis P. Garvan

Francis P. Garvan was president of the Chemical Foundation until his death in 1937. The Foundation, conceived by Garvan, facilitated U.S. manufacturers using patents, trademarks, copyrights, and contracts that had been seized by the Alien Property Custodian following World War I. Shareholders received no dividend; the president and vice-president worked without salary. The Foundation contributed 72% of its income of $8.6 million to the support of scientific research and education, of the *J. Chem. Ed.* for its first nine years, and of the microbiological journal, *Stain Technology*. It helped provide chemical references to libraries throughout the nation and supported the commission on Standardization of Biological Stains. The Garvans contributed an estimated one million dollars to help the Foundation and to finance, as a memorial to their daughter, the Prize Essay Contest sponsored by the ACS for seven years.

Garvan, a Yale graduate, received his LLD from New York Law School in 1889. After a distinguished law career he became Director of the Bureau of Investigations of the Office of the Alien Property Custodian, which led to his appointment as Alien Property Custodian and thus to his scientific career (*3*). Garvan is to date the only layman in the history of the ACS to receive its Priestley Medal (*4*).

Year	Name	Topic	Institution
1937	Emma Perry Carr	Hydrocarbon Structure by Far UV	Mt. Holyoke College
1940	Mary Engle Pennington	Food Chemistry	Private Consultant
1942	Florence B. Seibert	Chemistry of Tuberculosis	University of Pennsylvania
1946	Icie Macy-Hoobler	Nutrition Chemistry	Children's Fund of Michigan
1947	Mary Lura Sherrill	Molecular Structure	Mt. Holyoke College
1948	Gerty T. Cori	Enzymatic Synthesis and Reactions	Washington University, School of Medicine
1949	Agnes Fay Morgan	Chemistry of Vitamins	University of California at Berkeley
1950	Pauline Beery Mack	Calcium Chemistry of Bone	Penn State University
1951	Katherine Blodgett	Monomolecular Films	General Electric Research Lab.
1952	Gladys Emerson	Chemistry of Vitamin E	Merck Institute for Therapeutic Research
1953	Leonora Neuffer Bilger	Asymmetric Nitrogen Compounds	University of Hawaii
1954	Betty Sullivan	Cereal Chemistry	Russell-Miller Milling Co., Vice-President
1955	Grace Medes	Discovery and Study of Tyrosinosis	Lankenau Hospital Research Institute
1956	Allene R. Jeanes	Fundamental Research on Dextran	Northern Utilization Research Branch, USDA
1957	Lucy W. Pickett	Vacuum Ultraviolet Pioneer	Mt. Holyoke College
1958	Arda A. Green	Purification of Enzymes	Johns Hopkins University
1959	Dorothy V. Nightingale	Organic Synthetic Reactions	University of Missouri
1960	Mary L. Caldwell	Crystalline Enzyme Preparation	Columbia University
1961	Sarah Ratner	Protein Production Controlling Enzymes	Public Health Research Institute, New York, N.Y.
1962	Helen M. Dyer	Experimental Carcinogenesis Mechanisms	National Cancer Institute
1963	Mildred Cohn	Oxygen 18 Enzyme Mechanism Studies	Johnson Foundation, University of Pennsylvania
1964	Birgit Vennesland	Enzymic Hydrogen Transfer Studies	University of Chicago
1965	Gertrude Perlmann	Studies of Protein Structure	Rockefeller Institute
1966	Mary L. Petermann	Cellular Chemistry	Sloan-Kettering Institute of Cancer Research
1967	Marjorie J. Vold	Theoretical Models of Colloids	University of Southern California
1968	Gertrude B. Elion	Drugs for Chemotherapy	Burroughs-Wellcome and Co.
1969	Sofia Simmonds	Bacteria Amino Acid Metabolism	Yale University
1970	Ruth Benerito	Studies on Cellulose Properties	Southern Utilization R&D Division, USDA
1971	Mary Fieser	Chemical Literature	Harvard University
1972	Jeannine M. Shreeve	Inorganic Fluorine Compounds	University of Idaho
1973	Mary L. Good	Mössbauer Spectroscopy	Louisiana State University
1974	Joyce Kaufman	Quantum Calculations of Drug Action	Johns Hopkins University
1975	Marjorie C. Caserio	Physical Organic Chemistry	University of California—Irvine
1976	Isabella L. Karle	Crystallography	Naval Research Lab., Washington, D.C.

Index

READ ALSO . . .

A Devotion to Their Science: Pioneer Women of Radioactivity

Marelene F. Rayner-Canham & Geoffrey W. Rayner-Canham, editors and senior authors
Biographical essays on 23 women who worked in atomic science
during the first two decades of the twentieth century.
Copublished by the Chemical Heritage Foundation and McGill-Queen's University Press;
CHF has world rights except in Canada.
1997. 280 pp. Cloth, ISBN 0-941901-16-5, $55.00. Paper, ISBN 0-941901-15-7, $19.95

TOOLS FOR TEACHERS . . .

Chemical Achievers: The Human Face of the Chemical Sciences

Mary Ellen Bowden
This lively and comprehensive collection of photographs and
biographies shows the human side of science.
1997. 187 pp, paper, ISBN 0-941901-12-2, $20.00
Teachers: Ship to your school's address and pay only $10.00

Introducing the Chemical Sciences: A CHF Reading List

Reference works; histories of science and technology; histories of
the chemical sciences and industries; journals; more.
1998. 24 pp, paper, ISBN 0-941901-18-1, $7.50
($6.00 each for orders of 10 copies or more)

History of Chemistry Wallcharts

These timelines summarize key achievements in specific areas of chemistry.
• The Origins of Organic Chemistry 1800–1900
• Chemical Atomic and Molecular Theory 1800–1900
• Modern Atomic and Molecular Theory 1900–1960
• Industrial Chemistry 1800–1900 • Industrial Chemistry 1900–1960
• Analytical Chemistry 1800–1900

Published by the Royal Society of Chemistry; CHF is the North American distributor.
All posters are 33 × 23.25 folded to 8.25 × 11.5. Single-sided.
$12.00 each or $54.00 for set of six (25% off the full price!)

CONTACT:
Chemical Heritage Foundation, Books, Dept. WIC
315 Chestnut Street, Philadelphia, PA 19106-2702
Phone: 215-925-2222; Fax: 215-925-1954